Joseph Kendall Freitag

Architectural Engineering

With Special Reference to High Building Construction

Joseph Kendall Freitag

Architectural Engineering
With Special Reference to High Building Construction

ISBN/EAN: 9783743687042

Printed in Europe, USA, Canada, Australia, Japan

Cover: Foto ©berggeist007 / pixelio.de

More available books at **www.hansebooks.com**

ARCHITECTURAL ENGINEERING:

WITH SPECIAL REFERENCE TO

HIGH BUILDING CONSTRUCTION,

INCLUDING MANY EXAMPLES OF

CHICAGO OFFICE BUILDINGS.

BY

JOSEPH KENDALL FREITAG, B.S., C.E.

FIRST EDITION.
FIRST THOUSAND.

NEW YORK:
JOHN WILEY & SONS.
LONDON: CHAPMAN & HALL, LIMITED.
1895.

Copyright, 1895,
By
JOSEPH K. FREITAG.

ROBERT DRUMMOND, ELECTROTYPER AND PRINTER, NEW YORK.

PREFACE.

THE author has attempted, in the following pages, to define and illustrate, in a manner as practicable as possible, such of the fundamental principles in the design of the modern high building as may prove useful to architects and engineers alike.

While the technical press of the country has devoted considerable attention to many of the individual subjects here considered, yet the realization of a want of collective data on the subject of Architectural Engineering has induced the writer to present this volume.

As more and more of the principles of construction are being added to the curricula of our architectural schools, and as many of our engineering students are adopting building construction as a specialty, it is hoped that this effort will serve to unite still more closely the work of the one with that of the other.

The author would mention the efforts of one highly esteemed and dearly beloved in the engineering profession, Mr. E. L. Corthell, who has been striving for several years to see the two professions united by establishing an International Institute of Engineers and Architects, as well as a technical School of Architecture and Engineering at the new University of Chicago. The writer would also acknowledge the warm interest displayed in this work by his former professor of engineering, Prof. C. E. Greene, of the University of Michigan.

The following chapters are arranged in the order in which the calculations for such structural work must proceed, starting with the load-bearing floor system, thence through the successive stages to the foundations. The latter would seem to require the first attention ; but as they are the last to be calculated, being dependent on all other considerations, they have here been placed last. The illustrations and examples given have been largely obtained through the courtesy of the architects of the respective buildings. An endeavor has been made to present only the most practical methods.

JOSEPH KENDALL FREITAG.

CHICAGO, MAY, 1895.

CONTENTS.

CHAPTER I.
INTRODUCTORY 1

CHAPTER II.
FIRE PROTECTION 9

CHAPTER III.
SKELETON CONSTRUCTION—EXAMPLES—ERECTION, ETC 24

CHAPTER IV.
FLOORS AND FLOOR FRAMING 54

CHAPTER V.
EXTERIOR WALLS—PIERS 88

CHAPTER VI.
SPANDRELS AND SPANDREL SECTIONS—BAY WINDOWS 100

CHAPTER VII.
COLUMNS 113

CHAPTER VIII.
WIND BRACING 136

CHAPTER IX.

PARTITIONS—ROOFS—MISCELLANEOUS........................ 163

CHAPTER X.

FOUNDATIONS.. 171

CHAPTER XI.

UNIT-STRAINS—SPECIFICATIONS 201

CHAPTER XII.

BUILDING LAWS... 216

LIST OF ILLUSTRATIONS.

FIG.		PAGE
1.	Reliance Building, Chicago.	17
2.	Arrangement for Pipe-space in Halls.	22
3.	Chicago Stock Exchange Building. Perspective.	25
4.	Chicago Stock Exchange Building. Basement Plan.	27
5.	Chicago Stock Exchange Building. Ground Floor Plan.	28
6.	Chicago Stock Exchange Building. Typical Office Floor Plan.	29
7.	Marquette Building, Chicago. Perspective.	30
8.	Marquette Building, Chicago. Typical Office Floor Plan.	31
9.	Reliance Building. Typical Office Floor Plan.	32
10.	Masonic Temple, Chicago.	34
11.	New York Life Insurance Building. Perspective.	35
12.	New York Life Insurance Building. Plan of Banking Floor.	36
13.	New York Life Insurance Building. Typical Office Floor Plan.	37
14.	Fort Dearborn Building. Perspective.	38
15.	Fort Dearborn Building. Typical Office Floor Plan.	40
16.	Champlain Building. Typical Office Floor Plan.	41
17.	Old Colony Building. Perspective.	42
18.	Typical Framing Plan of Fort Dearborn Building.	43
19.	Typical Framing Plan of Reliance Building.	44
20.	Reliance Building during Construction.	48
21.	Reliance Building during Construction.	49
22.	Brick Arch Construction.	55
23.	Corrugated Iron Arch.	55
24.	Tile Arch used in Equitable Building, Chicago (1872).	55
25.	Tile Arch used in Montauk Building, Chicago (1881).	56
26.	Tile Arch used in Home Insurance Building, Chicago (1884).	56
27.	Arch showing Tile Filling Blocks used in Woman's Temple, Chicago.	57

FIG.		PAGE
28.	Panelled Beam, Fire-proofed	58
29.	Fire-proofed Girder	58
30.	The Lee Flat Arch	59
31.	The Johnson Type of Flat Arch	61
32.	The Austria Tile Arch	65
33.	The Melan Arch, Short Span	67
34.	The Melan Arch, Long Span	67
35.	Arch of Metal Straps and Concrete	69
36.	Arch of Wire and Concrete, Panelled Soffit	69
37.	Arch of Wire and Concrete, Flush Soffit	70
38.	Elliptical Concrete Arch	71
39.	Segmental Tile Arch	72
40.	Segmental Tile Arch used in Sibley Warehouse, Chicago	72
41.	Standard Connection-angles	85
42.	Standard Connection-angles	86
43.	Isometrical View of Connection of Floor-beam to Girder	87
44.	Detail of Terra-cotta Front. Reliance Building	93
45.	Section through Wall at Main Entrance to Masonic Temple	94
46.	Detail of Corner Pier for Reliance Building	97
47.	Detail of Wall Girders in Reliance Building	98
48.	Diagram of Thickness of Walls for Buildings Devoted to Sale and Storage of Merchandise	99
49.	Diagram of Thickness of Walls for Hotels and Office Buildings other than Skeleton Construction	99
50.	Diagram of Thickness of Walls for Office Buildings carrying Wall Weight only	99
51.	Spandrel Section. Ashland Block	101
52.	Spandrel Section. Reliance Building	101
53.	Connection of Cast Mullions. Reliance Building	101
54.	Spandrel Section, 11th floor. Fort Dearborn Building	102
55.	Spandrel Section, 12th floor. Fort Dearborn Building	103
56.	Spandrel Section, 1st floor. Fort Dearborn Building	103
57.	Spandrel Section, Roof and Cornice. Fort Dearborn Building	104
58.	Spandrel Section. Marquette Building	105
59.	Spandrel Section. Marshall Field Building	106
60.	Spandrel Section. Marshall Field Building	106
61.	Spandrel Section through Court Wall of Marshall Field Building	107
62.	Spandrel Section through Typical Court Wall	108

LIST OF ILLUSTRATIONS.

FIG.		PAGE
63.	Spandrel Section through Bay Window. Masonic Temple	109
64.	Spandrel Section at Bottom of Bay Window. Masonic Temple	109
65.	Half Plan of Metal-work in Bay Window. Reliance Building	110
66.	Half Plan through Bay-window Walls. Reliance Building	110
67.	Spandrel Section through Centre of Bay. Reliance Building	111
68.	Spandrel Section at Side of Bay. Reliance Building	111
69.	Floor and Ceiling Supports in Bay Window. Reliance Building	112
70.	Details of Joints for Cast Columns	114
71.	Detail of Larimer Column	122
72.	Detail of Larimer Column	122
73.	Detail of Gray Column and Connecting Girders	125
74.	Detail of Phœnix Column	125
75.	Detail of Z-bar Column. Monadnock Building	127
76.	Detail of Phœnix Column	128
77.	Detail of Phœnix Column used in Old Colony Building	129
78.	Detail of Box Column	129
79.	Section of Z-bar Column used in "The Fair" Building	130
80.	Method of Fire-proofing Phœnix Column	134
81.	Method of Fire-proofing Box Column	134
82.	Method of Fire-proofing Z-bar Column	134
83.	Method of Fire-proofing Columns in Monadnock Building	134
84.	Diagram of Wind Bracing by means of Sway-rods	139
85.	Diagram of Wind Bracing by means of Sway-rods	139
86.	Diagram of Wind Bracing by means of Portals	139
87.	Diagram of Wind Bracing by means of Knee-braces	139
88.	Figure showing Analysis of Sway-rod Bracing	141
89.	Figure showing Typical Sway-rod Bracing	143
90.	Wind Bracing used in Masonic Temple	144
91.	Floor Plan of Venetian Building	144
92.	Wind Bracing in Venetian Building	145
93.	Detail of Channel-struts. Venetian Building	146
94.	Detail of Cast Blocks. Venetian Building	146
95.	Partial Cross-section of Venetian Building	147
96.	Figure showing Analysis of Portal Bracing	149
97.	Portal-strut used in Monadnock Building	151
98.	Cross-section showing Portals in Old Colony Building	151
99.	Detail of Portal in Old Colony Building	152
100.	Figure showing Analysis of Knee-bracing	153

FIG.		PAGE
101.	Detail of Knee-bracing used in Isabella Building	154
102.	Channel-struts and Gussets used in Exterior Walls of Fort Dearborn Building	155
103.	Detail of Column Joint in Pabst Building, Milwaukee	158
104.	Detail of Column Splice in Reliance Building	159
105.	Detail of Book Tile	164
106.	Hall and Main Entrance to Marquette Building	166
107.	Hall and Main Entrance to New York Life Insurance Building	167
108.	Hall and Main Entrance to Fort Dearborn Building	168
109.	Rail Footing	174
110.	Masonry Footing	174
111.	Beam and Rail Footing	181
112.	Beam Footing used in Marquette Building	184
113.	Double Footing used in Marquette Building	184
114.	Plan of Cantilever Footing	186
115.	Elevation of Cantilever Footing	186
116.	Line of Flexure for Continuous Girder	187
117.	Figure showing Analysis of Cantilever Footing	187
118.	Figure showing Analysis of Continuous Girder	189
119.	Plan of Foundations. Manhattan Life Insurance Building, New York	199
120.	Cross-section showing Foundations of Manhattan Life Insurance Building, New York	200

ARCHITECTURAL ENGINEERING.

CHAPTER I.

INTRODUCTORY.

AMONG the most noteworthy examples of Architectural Engineering in recent years, "Le Tour Eifel" stands unique —a most perfect expression of this recently coined term, signifying a complete union of the great art of architecture and the science of engineering. While universally accepted as distinctly an engineering feat, this tower possesses such perfect structural beauty that it may well lay claim to the eulogies of architectural critics—eulogies that should be all the more emphatic when we stop to consider how few and far between, in modern times, are the creations of the engineer that can, at the same time, appeal to the architectural artist or designer, as embodying the beauty of form with the excellence of construction; while the reverse may truly be said of modern architecture. For who may claim justly that our present architectural efforts are true, characteristic expressions of modern life, or reflections of the progress that has characterized our age—as classical architecture embodied classical life and mediæval architecture expressed mediævalism?

The science of engineering has, at least, been progressive, keeping pace with modern developments, while architecture, for the most part, has been stationary, content to copy the original form of a civilization whose substance has undergone ages of evolution. Hence arise the causes for the present antagonism in these two closely related professions. There should be none, but that there is, no one will deny. One of the most prominent engineers of the United States has been heard to characterize architects as "milliners," and their work as "millinery" or "gingerbread decoration"; while the architect, on his own little pedestal of pure art, scorns the engineer as incapable of producing the beautiful. There is, doubtless, partial justice in each of these criticisms; the architect's blind devotion to classic forms becoming as much of a hindrance to the practical aims of the engineer, as the barren stamp of utility, glaring from a purely engineering work, is an offence to the eye of the artistic designer. But the keynote has already been sounded for a more perfect union between these two professions, each of which is the necessary complement of the other.

Although the term architectural engineering has but recently sprung into use, the perfect union of the two arts is as old as the arts themselves. Pyramids, obelisks, temples, palaces, and sepulchres, all show that the architects of early days were the engineers as well. Vitruvius, the only ancient whose ideas on architecture have been preserved for us, established three qualities as indispensable in a perfect building: stability, utility, and beauty,—the first two of which certainly lie within the range of the science of engineering. As a proof that those early architects were governed by the laws of Vitruvius, we have but to look upon the pyramids of Egypt, the vast monoliths of Rome, the temples of Sicily, or the massive Parthenon. Their graceful proportions and harmony of design have for centuries made of the architect an admiring copyist, while their massiveness and stability suggest to the engineer the possibilities of human power. Take for example one of the largest pyramids not far from the city of Cairo. This rough, awe-inspiring mass of masonry covers 11 acres of the sands of the Nile, while its height is but little less than our Washington Monument, or nearly 500 feet. Again, consider the temple of Babylon, 660 feet in height, built of blocks of stone 20 feet long, used in a brick-like fashion, some of them being 15 feet broad and 7 feet thick; or the massive remains of an Egyptian temple, the walls of which were found to be 24 feet thick; while at the gates of Thebes the foundation walls were 50 feet thick and perfectly solid.

The ethnologist tells of an age of clay, then stone in the rough and, later, polished; an age of bronze, then iron; and now we add steel and the newer materials. Architecture, as represented in the temples, tombs, palaces, and habitations of man, has always been, like literature, the surest indication of the customs, arts, and needs of the people who produced it—in fact, a perfect reflection of the civilization in which it is found. "Cain, the son of Adam, builded a city,"—the rude mud hut or the flimsy structure of reeds serving as man's habitation in primitive times, imitating the nests of birds, of which modifications still exist in China and other Eastern countries, as well as in many parts of dark Africa. The later days of clay and straw and then burned brick were succeeded by the age of stone, reaching such a height of excellence in the works of the Greeks and Romans, and the castles and cathedrals of the middle ages. The temple of Solomon, rebuilt by Herod at Jerusalem, was, so the Bible states, 46 years in erection, with stones 46 feet long, 21 feet high, and 14 feet thick, while some were of the great length of 82 feet. Would it not tax the ingenuity of an engineer in our own advanced age to handle such masses of stone? Architecture and engineering certainly worked in harmony in these examples, which must ever rank with the greatest creations of man.

Now, with the hurrying strides of civilization, comes a demand for a cheaper and quicker construction, a medium capable of being more easily handled than the huge blocks of stone of early ages; while the principles of statics and the economics of construction present themselves with ever increasing clamor for solution and application, until we boast that our age is one of specialties, involving an exactness hitherto unknown in the observance of all the laws of nature formulated, as they are, into exact sciences.

It was but natural, in the examples we have considered, that architecture should go hand in hand with engineering, for the architect was the engineer, employing rule of thumb methods, to be sure, and knowing little of the laws of statics or dynamics. Indeed it was not till the thirteenth century that the solution of the theory of arches and vaults was attempted. Old, old indeed, is the relation of friendship that has existed between the naturally allied arts of architecture and engineering—a mutual bond, which will, we believe, give us still more perfect examples of the strength and beauty that architectural engineering makes possible: *architectural*, in reference to the expression and beauty of the edifice—*engineering* (perhaps partially, if not wholly, hidden from the eye), in construction, durability, and magnitude that result from the possibilities which open up before the mind accustomed to dealing with the matter and forces of nature, and adapting them to the ever-increasing wants of an exacting public. The materials of nature assume higher and higher planes in the fulfilment of man's needs, as he constantly overcomes more of the natural destructive elements and agencies by applying himself with scrupulous exactness to every detail of work. Considering the present tendency to specialization, it seems absurd to suppose that the architect may eventually be employed simply as an ornamental draughtsman by the engineer, or that the engineer may become subservient to the architect. Either profession is too noble and comprehensive in itself to permit of such absorption. It is but a natural prejudice to give first importance to one's own branch of work; and, indeed, the engineer quite justly claims a prerogative, since upon the accuracy of his calculations depend the stability of the structure and the safety of the tenants. But, on the other hand, one cannot severely censure the architect for ridiculing such work as many of our best engineers send forth, as devoid of beauty or even harmony of line. It is apparent, therefore, that the truest expression of our life and civilization must be found in a more perfect harmony of these two professions. The architect of early days was enabled by rule of thumb methods, good judgment, and a knowledge of past examples to produce the structures he built; but with the exactness of our professional work at the present time, and the multifold necessities of our comprehensive civilization, the architect who endeavors to compass the sphere of the trained engineer will find the longevity of Methuselah desirable for his education. Let the engineer know more of art and appreciate its value, and let the architect know as much as possible of construction and the laws of the forces of nature. But that either may fully grasp the details of both professions seems well-nigh impossible.

The architect has been accustomed to say that such a perfect union is impracticable, but the architectural critics of to-day are demanding it, as is shown by the following: "In art, as in nature, an organism is an assemblage of interdependent parts, of which the structure is determined by the function, and of which the form is an expression of the structure." Again: "That form is pleasing to good taste which shows and reveals its use. That form reveals the use most successfully whose surface and outlines and whose skeleton or frame speak for themselves, and are not obscured by misplaced ornament."

If these quotations from purely architectural critics, without reference to engineering, are to be given value, then surely a more rational union of excellent structural design on economic principles, with perfect architectural expression of the underlying organism, is not only possible but necessary for a proper reflection of our civilization. The people of our country have demanded "sky-scrapers," in accordance with the strong tendency to centralization. New problems have been created and new necessities imposed, and the engineer has come to the front with the steel and terra-cotta of the "Chicago construction," as the means of solution on his part; but it remains for the architect to give true expression and permanent form to what the engineer has evolved. It is from the union of the results obtained by a rational division of labor in the art of building that we hope for the perfect architecture of the present age.

It has been said that our civilization has demanded a medium of construction more in accord with the push and hurry and economy of our day than is found in the massive masonry construction; a substance combining the strength, durability, and adaptability required by the demands of commerce and rapid progress. In the architectural history of our own country we have not confined ourselves to any one material long enough to develop for it a unique, characteristic style of representation. Our architectural form has, rather, been a series of rapid changes. The refined and sober examples of our colonial forefathers rapidly gave way to the more ostentatious efforts of the jig-saw in our frame construction, and this period of the shingle and fretwork gave place to the rows of red brick, and later the brown-stone front, with all its attendant horrors in galvanized iron. Cast iron, too, has held its sway for its own little period, only to be displaced by its more refined and enduring successor, steel. Our present epoch has been characterized so often as one of steel and terra-cotta that the subject is becoming trite indeed ; but only through the combination of these materials have the huge frame-works which now mark our large American cities become possible. And as the "skyscraper" office buildings present interesting problems in architectural engineering, which are being constantly discussed in the technical press of to-day, some of the constructional points involved will be considered here with special reference to "Chicago construction," a construction which has almost universally been attributed to the skill of the *architect*, though in only too many cases the architect, who has designed, as it were, the sugar coat, to make the exterior palatable to the public, reaps all the reward for what the engineer has made possible.

The fact that in our large cities it is found most advantageous as to time and convenience for business transactions, to have our commercial headquarters and office-building district concentrated within a limited area, has caused the adoption of buildings numbering from 16 to over 20 stories. Increased floor-space must be obtained to realize on the investment, and it is evident that these tower-like structures must continue to increase in numbers, when we note the abnormally enhanced prices to which the value of land is rising. The American Surety Company in New York City might be mentioned as paying the sum of $1,500,000 in 1894 for a piece of property about 85 feet square, which would be at the rate of $8,000,000 per acre.

The continued development, however, of this central

zation of business operations is attended by many vexing difficulties, the attempted solution of which has caused a number of clauses of restriction to appear in the municipal building laws. Considerable discussion has been going on about the sanitary aspect of this question; the damp, unwholesome, and microbe-laden air which must lurk in the deep valleys or streets between mountainous structures on each side; the dark and uninviting offices of the lower stories, which would soon become vacant; and the congested condition of our sidewalks when our vertical carrying capacity is greater than our horizontal or street capacity—all are considerations of grave importance.

But that the *proper* regulation of building operations, with their attendant difficulties and future possibilities of development and style, may be successfully accomplished by general municipal *ordinances* is very doubtful. The building laws of many of the larger cities already prescribe a maximum height for all structures; but considering high buildings, *per se*, it is evident that it is not so much legislation limiting the possibilities of design that is needed, as it is laws compelling the appointment of competent engineers to supervise the designs, specifications, and execution of large buildings, and possibly a competent board of architects to pass on the proposed location of an extraordinarily high structure. It would be well if we adopted more of the European practice, giving harmonious appearance to our thoroughfares and considering the specific conditions of each new structure of monumental pretensions, instead of binding all through an inflexible law. Edifices fronting on parks or open spaces might then be treated in more heroic proportions than those of narrow by-ways, and the incongruous mixture of ups and downs, side by side, might give place to some semblance of harmony between neighbor and neighbor.

CHAPTER II.

FIRE PROTECTION.

BEFORE considering the details of skeleton construction it will be well to consider the general subject of fire-proofing, with its effectiveness and its limitation.

The total fire loss in the United States during the year 1894 was about $128,000,000, of which the insurance companies paid, as their share, some $81,000,000. This stupendous drain on the resources of the nation may be better appreciated if we consider that the full value of the pig iron production for the same year was about $75,000,000.

When to this fire loss we add the estimated amount necessary to maintain the fire departments, and to sustain the fire insurance companies, the grand total will exceed $175,000,000 annually.

If, then, it is true, as stated by underwriters that forty per cent of all fires are attributable to causes easily prevented, a proper treatment of the fire problem certainly becomes a very practical and economic inquiry.

The subject of proper fire protection is now recognized as a legitimate and important branch of engineering. It is no longer confined exclusively to endeavors to protect human life, but is greatly increasing in scope, demanding very careful thought from its economic standpoint as well. The old adage of an ounce of prevention being better than a pound of cure is slowly but surely demonstrating its truth as applied to the ravages of fire, as well as of disease, and the specialist who enters this broad field of research and improvement must meet causes and effects with a precision not less exact than does his medical brother. Conflagration has formerly been looked upon as an inevitable calamity, inflicted by a supernatural agency; and property-owners have been content, year after year, to pay enormous insurance rates, suffering with resignation the destruction of their property and the annihilation of their business. Add to these the loss of articles of peculiar associations,—heirlooms and treasures of art and science,—and the possibility of relief from this Damoclean sword of conflagration is a liberation indeed. And that this question of fire waste is being seriously considered in all its aspects and by all classes of society is shown by the widening facilities for the use of fire-proof construction. The realization of low prices in the building market has served to overthrow many of the hitherto unquestioned prejudices in regard to fire-proof construction, and the economy of such design as opposed to the fire-trap methods so long in vogue is now being daily emphasized by architects, engineers, and the technical press. And what is most gratifying is the fact that this economy is beginning to be appreciated not only by the owners of palatial office buildings, stores, and magnificent residences, but also by people of limited means, as is evidenced by the start already made in fire-proofing the ordinary city house, at a figure not exceeding the cost of present methods. It was found recently, in taking figures for a building in Philadelphia to cost $125,000, that a thoroughly fire-proofed construction would cost only 3.6 per cent more than the ordinary method of building. This increase would be compensated for in a very short time by the decreased insurance.

The tide has turned, and nothing can stay the flood of progress in this direction. The dawn of the twentieth century will undoubtedly see nearly all of our mercantile, manufacturing, and even dwelling houses, except those of the very cheapest description, built according to fire-resisting principles. Steel, the clay products, and cement or concrete are the materials of the future, permanent, fire-resisting, of ready adaptability, and of remarkably low cost. The fire-trap timber construction, threatening the exhaustion of our vast forestry resources, accompanied by its susceptibility to dampness, drought, heat, and cold, involving dry-rot, as shown by the collapse some years ago of a prominent hotel in Washington, must give way to new conditions, and further improvement in a field of such promise. The insurance burden will be gradually lightened, and human life be better protected.

While buildings *could* be erected with absolutely no inflammable material in their construction, there would still remain the furniture and property of the tenants to feed possible fire. This element of danger cannot be eliminated; and added to this are the dangers that come from without as well as from within. For as long as highly inflammable buildings surround even the most excellent of modern fire-proof structures the term is but mockery. Fire-proof structures must stand in fire-proof cities. Hence the word "fire-proof," as applied to modern structures, does not mean one that claims immunity from all danger of fire, for considerable woodwork must still be used in interiors, and the average contents are dangerous in the extreme; but it does claim to embody principles which have reduced the fire hazard, both interior and exterior, to a minimum, according to the best skill and judgment of the day. The term implies that all structural parts of the edifice must be formed entirely of non-combustible material, or material which will successfully withstand the injurious action of extreme heat. Following is the definition given in the new building ordinance of Chicago: "The term 'fire-proof construction' shall apply to all buildings in which all parts that carry weights or resist strains, and also all stairs and all elevator enclosures and their contents, are made entirely of incombustible material, and in which all metallic structural members are protected against the effects of fire by coverings of a material which must be entirely incombustible and a slow heat-conductor. The materials which shall be considered as fulfilling the conditions of fire-proof coverings are: First, brick; second, hollow tiles of burnt clay applied to the metal in a bed of mortar, and constructed in such manner that there shall be two air-spaces of at least three-fourths of an inch each by the width of the metal surface to be covered, within the said clay covering; third, porous terra-cotta, which shall be at least two inches thick, and shall also be applied direct to the metal in a bed of mortar; fourth, three layers of plastering on metal lath, so applied upon metal furring that there shall be a solid layer of mortar at least one-half inch thick between the metal to be covered and the metallic lath, and then two air-spaces of at least three-fourths of an inch in the clear between the first-mentioned layer of plastering and the outer surface of the finished covering."

There are many materials quite satisfactory as fire-proofing mediums for the constructional parts of a building, but the inventor has yet to supply an acceptable incombustible material for the interior finish. The best that can be done, at present, is to reduce the inflammable elements to a minimum, and endeavor to confine the fire by means of fire-proof floors and partitions, so that it may do no injury beyond the consumption of local woodwork and furnishings. This may be accomplished largely by means of floors of concrete or terra-cotta with I-beams, using mosaic or marble tile instead of wood flooring, partitions of plaster board, cement or metallic lath, or terra-cotta blocks, and bases and wainscoting of marble. The possibility of using frames and casings for doors and windows made either of metal or sheet metal over wood, and doors covered with sheet metal, seems but a question of short time in adding further efficiency to high-class fire-proof structures. A metal-covered door has lately been introduced in this country, giving a well-appearing, light, and incombustible contrivance, and serving as an effectual barrier against the spread of flames. The success that has attended the use of wire glass in skylights has also prompted the suggestion to reduce the exterior hazard by protecting all windows, which offer the most vulnerable points of attack, by using a plate glass with silvered or gilded wires imbedded therein, in graceful patterns or network, serving the purpose of additional fire protection, as well as architectural effect. The planning of the building, and the proper location and installation of the various power plants and mechanical features, also become vital problems in fire-proofing.

The success that has attended past efforts in this direction may be judged by such examples of fire as have been afforded in protected structures. The largest and most interesting of such tests of the new methods was the burning of the Chicago Athletic Club building while under construction. Though not entirely satisfactory as a test of present building methods, " this building furnishes an assurance that was lacking before—that the metal parts of a building if thoroughly protected by fire-proofing, properly put on, will safely withstand any ordinary conflagration, if the quantity of combustible materials the building contains is not greatly in excess of that which enters into the construction of the building itself."

This extract from the report of experts employed to investigate this fire and its effects, emphasizes two very important facts, namely, the danger of the indiscriminate use of combustible material not absolutely necessary in the construction, and second, the evident superiority of terra-cotta as a fire-proofing substance.

The above fire, which occurred on November 1, 1892, was the first case on record of a fire in a building intended to be fully fire-proof where the loss to the insurance companies was more than thirty per cent of its value. It is further stated in the report that "if the building had been completed, it would never have contained combustible material enough (or so distributed) to have produced sufficient heat to have done any considerable damage to the building by burning."

The fire in question was of very intense heat, inasmuch as a vast quantity of scaffolding, flooring, trim, etc., was collected in mass, preparatory to use; but, in spite of this, there seemed no reason for questioning the integrity and strength of the building, as a whole, after the fire, and no doubt existed that the fire-proofing around the columns saved them from utter collapse, because it remained in place until the fuel that had fed the flames was well-nigh exhausted. The result to the building included the entire destruction of all the interior finish, plastering, piping, and wiring, as well as parts of the elaborate front of Bedford stone and pressed brick. But the steel columns and beams were uninjured, except a few of the latter where unprotected; and the tile arches, built after the end construction method, were almost uninjured, in spite of the combined action of great heat and frequent applications of cold water.

It is not advocated that fire-proofing as efficient (or inefficient when the preservation of human life is considered) as the foregoing example is sufficient for present needs—it certainly is not. But certain underlying facts have been clearly proved by this test, and taking these essential points as a basis, and using the utmost care and judgment in the matter of details. it must be admitted that the use of terra-cotta, as seen in the better examples of recent fire-proof buildings, goes a long way in the successful solution of one of the most important problems of modern times.

The method of fire-proofing now employed consists of a vital skeleton or frame-work of wrought iron or mild steel, enclosed in a continuous sheathing of terra-cotta. Every square inch of the metal-work must be protected by means of the various shapes made by the terra-cotta companies, thus avoiding all direct transfer of heat.

Hollow tile as a building material was first introduced in the United States in 1871, shortly after the great Chicago fire. Its first use was for floor arches, to replace the old brick arch method. Terra-cotta was a direct outcome from conditions imposed by the increased height, and hence weight, of a rapidly developing architectural construction, and its necessity was doubtless made more apparent by the great object lesson afforded by Chicago's disastrous conflagration. A substance was necessary to replace the heavy masses of masonry which constituted the fire-proofing at that date, both in the exterior walls and in floor arches, and the peculiar advantages of terra-cotta caused it to undergo many improvements in rapid succession, effecting not only its use in floor construction and column and beam protection, but adapting it to the needs of a lighter and more rapid construction throughout. Its attendant reduction in weight, its great fire-resisting qualities, its peculiar adaptability to all conditions of position and form, its susceptibility to modelling, and its readiness of manufacture in shapes convenient for transportation and erection, soon caused it to win favor both for its artistic possibilities and its enduring qualities, through which it becomes one of our most valuable constructive media.

First used in interior work only, it soon appeared in belt courses, sills, caps, ornamental panels and modelled work in the hard-finished terra-cotta, until to-day its use is almost more general than stone, appearing in entire fronts, as a bold-faced impersonation of solidity itself.

The field of architectural expression in terra-cotta has recently been widened to a still more remarkable degree by the successful completion in *enamelled* terra-cotta of the façades of the Reliance Building, Chicago, supplied by the Northwestern Terra-Cotta Co. (see Fig. 1). Should this material successfully withstand our severe climatic changes, and undergo the same course of rapid improvement as did the ordinary terra-cotta used in exteriors, a vast field for more extensive coloring effects would then be opened up to the architect who strives to create "a thing of beauty forever" in the smoke and soot-laden air of our American cities. The underlying idea of enamelled exteriors is, of course, that they may be readily washed down and cleansed of the soot which so soon destroys any attempts at light coloring.

With this general review of the fire problem, and terra-cotta as a weapon of defence, it becomes evident that a fire-proof structure must possess:

1. General excellence of design.
2. All floors of fire-proof construction.
3. All columns of masonry or steel, protected from fire.
4. All outside piers and walls of masonry or steel, protected from fire.
5. All partitions and furring of fire-proof construction.

There are three methods of general design advocated at the present time as means of reducing the fire risk—the "slow burning construction," the so-called "mill construction," and the still more effectual "fire-proof construction." The term "slow burning construction" is applied to build-

FIG. 1.—The Reliance Building. D. H. Burnham & Co., architects.

ings in which the structural members, carrying the floor and roof loads, are made of combustible material, but protected throughout from injury by fire, by means of coverings of incombustible, non-heat-conducting materials. Thus the wooden floor joists are protected on the under side by a single covering of plaster on metal lath, while a thickness of 1½ inches of mortar or incombustible deadening is required above the joists. Columns, if of oak, with a sectional area of 100 square inches or over, need not have special fire-proof coverings. Partitions and elevator enclosures must be wholly of incombustible material, and no wood furring is allowed.

Buildings of "mill construction" are those in which all floor and roof joists and girders have a sectional area of at least 72 square inches, with a solid timber flooring not less than 3¾ inches in thickness. Columns of wood need not be protected, but they should have a sectional area of at least 100 square inches. Partitions and elevator enclosures are of incombustible material, and no wooden furring or lathing is used.

"Fire-proof construction" has already been defined. The two types first mentioned do not, then, depend on the use of materials wholly incombustible, but rather on the judicious design and careful use of ordinary building materials, the aim being to provide structures so open and free from fire-lurking corners that they may offer no obstacles to a speedy suppression of the conflagration. These types are peculiarly adapted to large mills, warehouses, and the like.

The scientific fire-proofing of a building does not consist in a proper selection of materials alone, for a structure may be reasonably secure against accidental fire, or the extension of fire, even when built of combustible materials; nor does it lie merely in guarding against the causes of fire. It can be secured only by a thorough acquaintance with all the general features and minutest details of all kinds of structures, and by a quick perception "for the numerous elements of danger that are constantly creeping into modern systems of buildings." The plan must be carefully studied to secure means of cutting off communication between floor and floor, and between and around dangerous sources, isolating, if possible, all stairways and elevator-shafts by means of fire-resisting walls, and confining all power and mechanical plants in such a way that there can be no possible means of fire extension. It is true that most high office buildings do not possess the isolated stair-well or elevator-shaft, but if they do not, great care must be taken in making the halls and corridors of more than ordinary security. They will still be the means for a rapid distribution of smoke from floor to floor, and thus make the danger from suffocation assume an importance equal to that of fire. This threatening possibility has not yet verified itself, and it is to be hoped that it will be denied the opportunity.

No less important is the cutting off of all communication between pipe- and air-passages. Piping and passages of all kinds should be carefully considered as a part of the fundamental design, for they not only become great eye-sores from their exposed positions in offices, but they also serve to make many of our fire-proofing endeavors quite useless.

The architect or engineer must finally be well informed in regard to the details and varied uses of approved fire-proofing materials. These must include terra-cotta in all the different shapes made by the terra-cotta companies, cement, concrete, fire-brick, asbestos, mackolite, etc. A judicious and economic use of all these materials is necessary, so that the most practicable form may be chosen to secure the desired end.

Some of these important minutiæ may properly receive detailed attention, when we remember that the strength of a structure is gauged by its weakest point.

The metal columns, for example, are properly figured for their safe dimensions, but from this step on they are apt to become a bug-bear to both architect and owner, the former desiring to reduce their size to a minimum on account of appearance, while the latter considers that they deprive him of the revenue of just so much floor-space. Any measures are therefore adopted to reduce their size. First, the various waste-, heat-, and supply-pipes are run up alongside the columns from floor to floor. For the passage of these pipes openings must be made in the tile floor arches, which, in the rush of building operations, may never be properly filled up again. These openings come *inside* the line of the fire-proof slabs of the column, thus forming one long continuous flue from basement to roof. The finished line of the fire-proofed and plastered column is often not more than 2 in. from the extreme points of the metal-work, and then, deducting $\frac{1}{2}$ in. or $\frac{3}{4}$ in. for plaster, little enough is left for the fire-proofing proper. The various pipes before mentioned will very often project even farther than the column itself, thereby tempting the fire-proofer to trim and shave till the original *little* has become still *less*.

In the Athletic Club Building fire some of these points were illustrated with glaring prominence. A steel framework and fire-proof covering having been used as the main elements of construction, further consideration of fire hazards were apparently slighted. In no case did the fire-proofing extend more than 2 in. from the outermost edge of the ironwork, while *wooden* nailing-strips were embedded in the tile at intervals of about 3 ft. starting from the floor (a 4-in. face exposed), making successively 3 ft. of tile and 4 in. of wood. These nailing-strips were employed as grounds for the panelled oak wainscoting, and a further error was made in leaving an air-space behind this panelling, with no "back" plastering. The ceiling also left an air-space, due to 1-in. raised nailing-strips.

As a matter of course the wooden grounds around the column burned out, letting the fire-proofing fall in 3-foot sections. It so happened that but two columns were badly bent by the intense heat, but who can say what the stability of those re-used unbent columns really is? Were they cooled slowly, or suddenly by the application of streams of water, and thus rendered brittle, and were they heated unevenly, thus causing great strain in the material on but one side of the column? What was the amount of expansion and contraction? No experiments could be made with reasonable economy and safety to satisfy these queries, leaving the present state of the building an uncertain conjecture.

The proper installation and distribution of the mechanical features in a modern office building have been given considerable attention by John M. Carrére (see Eng. Mag., October, 1892), and the system proposed by him will undoubtedly add greatly to the efficiency of fire-proofing, and remedy many of the weak details just considered. In order to avoid chases, or continuous flues, the lowering of the hall ceilings is suggested, "thereby obtaining a horizontal space under the floors of the halls at each story, lined and fire-proofed, where all the mechanical features except steam heat can be placed" (see Fig. 2). An arrangement of this character would certainly possess many great advantages—it would always be accessible for repairs, easy of connection with all offices, and would serve as a *safe* and at the same time hidden conduit for all wiring, piping, and ventilating air-ducts, either exhaust or indriven.

The additional expense would not be great either, and when its permanency is considered, never being affected by the moving of partitions, etc., as is now the case, it is surprising that such a system has not attained more general use.

At the ends of these horizontal ducts are vertical chases or ducts built solidly of fire-proof blocks or brick from cellar to roof, and connected at each floor with the hori-

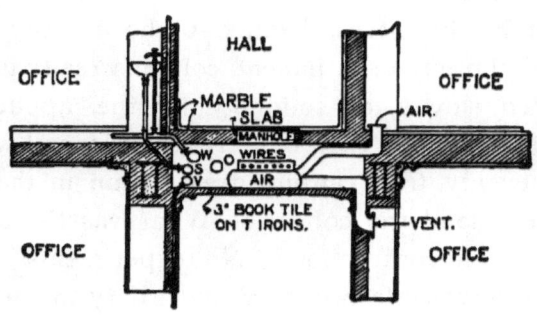

FIG. 2.

zontal leads, but still partitioned off at each floor with wire and plaster partitions, to prevent the spread of possible fire. All of the vertical risers could be placed in these chases, thus avoiding the unsightliness of pipes in the office space, or the necessity of placing such piping within the column space.

The growing importance of adequate fire protection may be judged from the care displayed in the encasing of the large girders at the new Tremont Temple in Boston. These girders carry columns of great load, and any warping tendency from great heat would be attended by most serious results. The steel girders were first surrounded by blocks of terra-cotta on all sides, and these blocks were then bound by iron bands. Over these blocks was stretched expanded metal lathing with a heavy coat of

Windsor cement. Iron furring was next placed on all sides to receive a second layer of expanded metal lath, on which was placed the finished plaster. The covering thus consisted of a dead air-space, terra-cotta blocks, a coating of cement, a second air-space, and an external coating of cement.

CHAPTER III.

SKELETON CONSTRUCTION—EXAMPLES, ERECTION, ETC.

MANY of the details which will be discussed in the following pages may be better appreciated in their relation to the whole subject if a few typical skeleton structures are examined. The scope of this outline will not permit of a discussion of the architectural problems involved in the design of a modern office building, hotel, or any of the structures which are now built according to skeleton methods. The points here considered are, rather, those of construction pure and simple. But the comprehensive view of the subject necessary to the architect or architectural engineer may only be obtained through an accurate knowledge of the manifold items which become a part of a successful plan. These accessories to the mere frame-work lie within the province of the engineer as well as of the architect, and here, as in the execution of the external expression of architectural engineering, a perfect harmony must exist between the two branches in the perfection of all mechanical details, if results are to be secured which may be looked upon as creditable to both professions.

The value of such accessories may be more fully realized when the self-sufficiency of a modern office building, containing all modern improvements, is considered. Electric light, the telephone, mail-chutes, and well-appointed toilet-rooms are already demanded as absolute necessities, while late examples provide telegraph and messenger service, cigar- and news-stands and barber-shops, besides restaurants and cafés in the basements. It is true that many of these factors would seem to have little bearing on the duties of the engineer, and yet it was just such conditions, imposed on the designer of the foundations of office buildings, that produced the successful development of the so-called raft or floating foundations, in order that the basements might be unencumbered by the large pyramidal

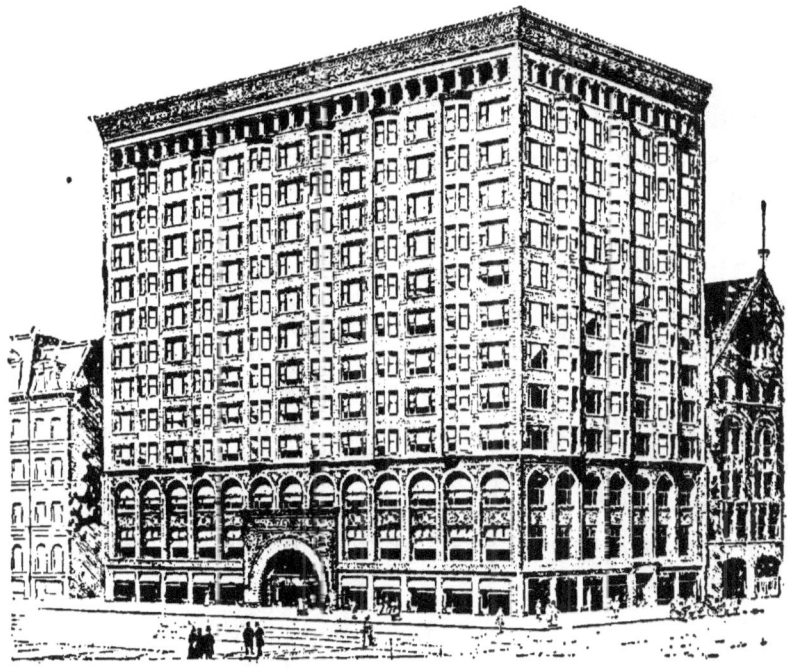

FIG. 3.—Chicago Stock Exchange. Adler & Sullivan, architects.

masses of stone previously used as footings, and the basement space might be added to the available renting area, or be used for the mechanical plants. The rigid economy of floor space which is demanded may only be obtained by careful attention to the most advantageous uses to which the different floors and rooms in the structure may be put.

Some examples of office buildings recently constructed in Chicago will here be given.

THE CHICAGO STOCK EXCHANGE.

A perspective of this building by Adler & Sullivan, architects, is shown in Fig. 3. The façades are constructed of a yellow-drab terra-cotta, with white enamelled brick in the interior court.

Fig. 4 shows the basement plan, containing the boiler- and engine-rooms, restaurants, etc.

Fig. 5 is a plan of the ground floor, showing the entrance vestibules, elevators, store areas, etc.

Fig. 6 gives a plan of the arrangement of the offices, etc., on the sixth floor. The toilet-rooms, barber-shop, vent spaces, and the arrangement of the lighting courts are plainly shown.

THE MARQUETTE BUILDING.

This office building (see Fig. 7), designed by Messrs. Holabird & Roche, architects, has but just been completed. The exterior walls are built mainly of a dark red brick, with terra-cotta base, cornice, and trimmings.

A typical floorplan, showing possible sub-divisions, is given in Fig. 8. Many of the floors in the larger office buildings are never subdivided until rented, in order that the arrangement of offices may be made to suit the tenant.

RELIANCE BUILDING.

Fig. 9 gives a typical floor plan of this building by D. H. Burnham & Co., architects. This arrangement of offices is intended for rooms to be used in suites. The pipe space at the side of the elevators, and the space for counterweights behind the elevators, are plainly shown, as is the circular smoke flue.

The elevator accommodations in these various buildings may be seen on the plans. Rapid passenger and freight service must both be provided for, and the necessary space allowed for the hydraulic cylinders in the basement, as well as for the vertical counterweights. Beams must be

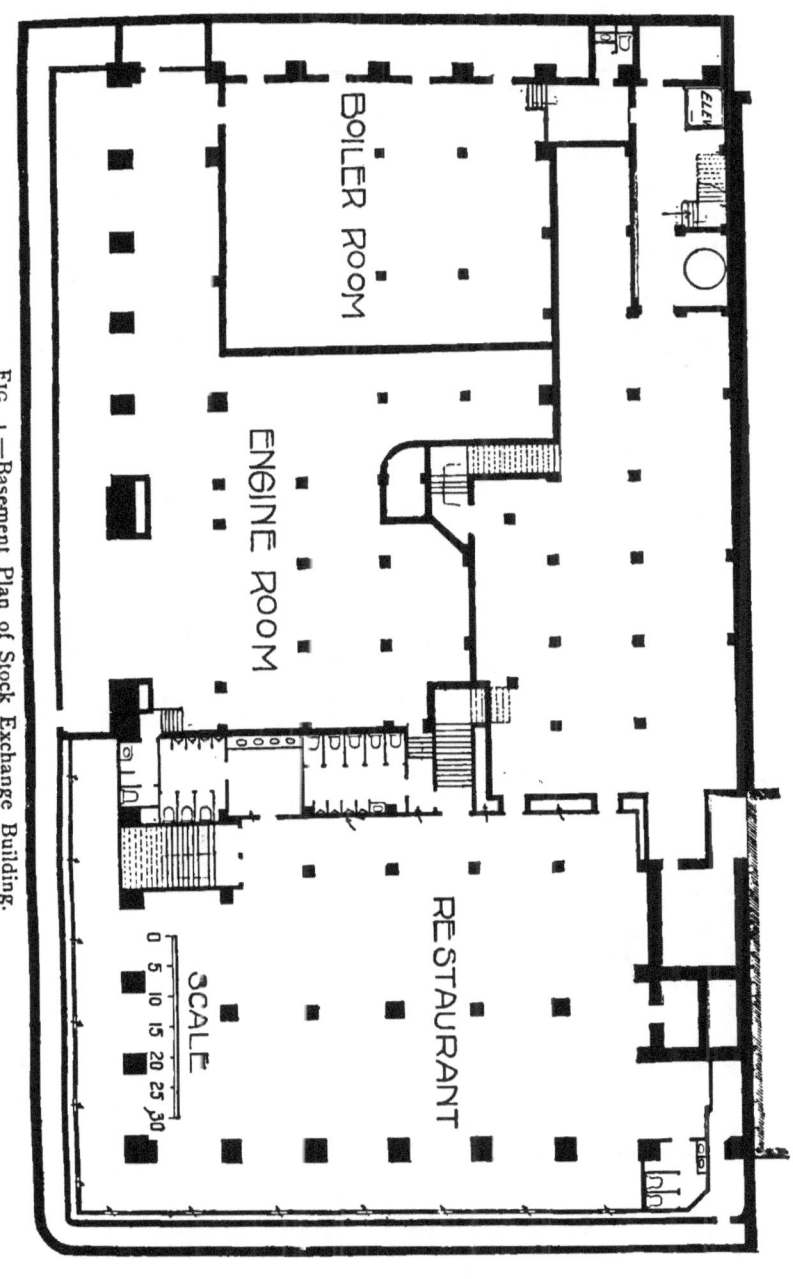

Fig. 4.—Basement Plan of Stock Exchange Building.

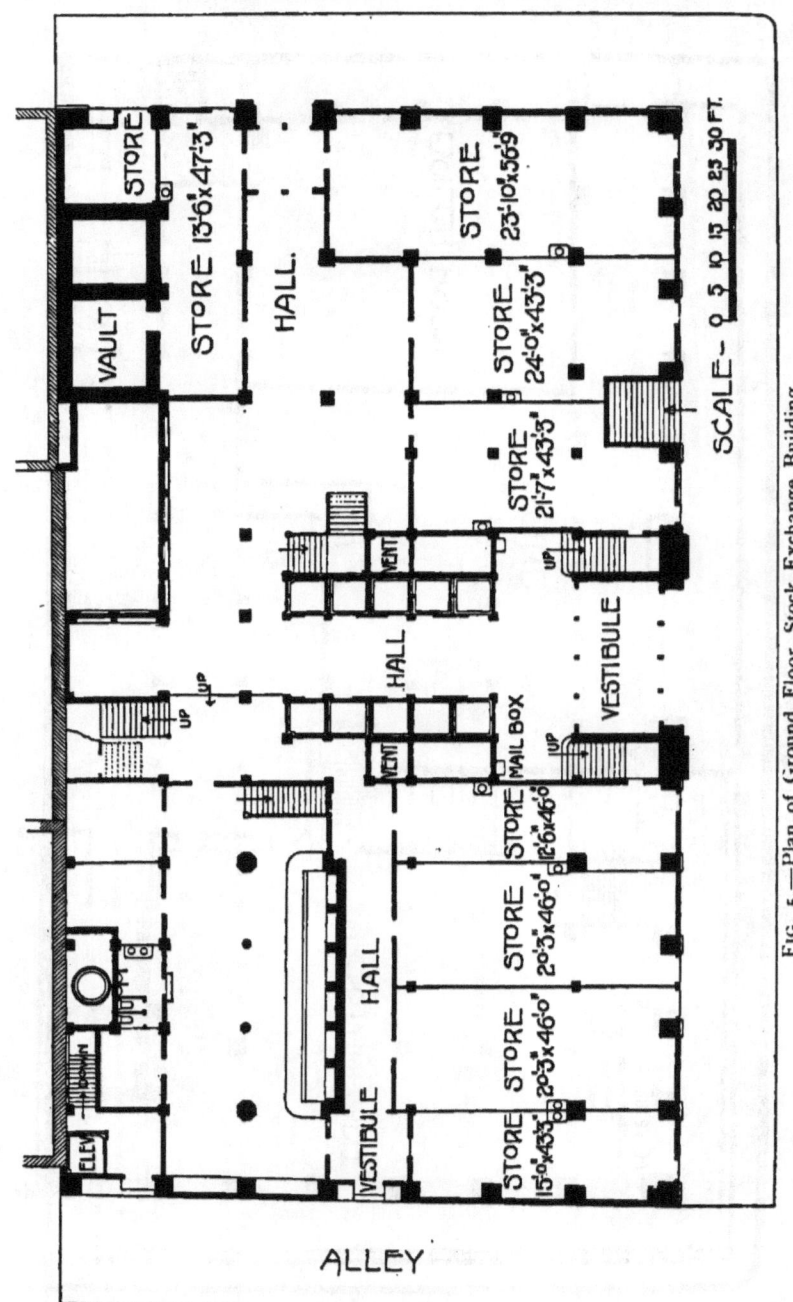

Fig. 5.—Plan of Ground Floor, Stock Exchange Building.

SKELETON CONSTRUCTION.   29

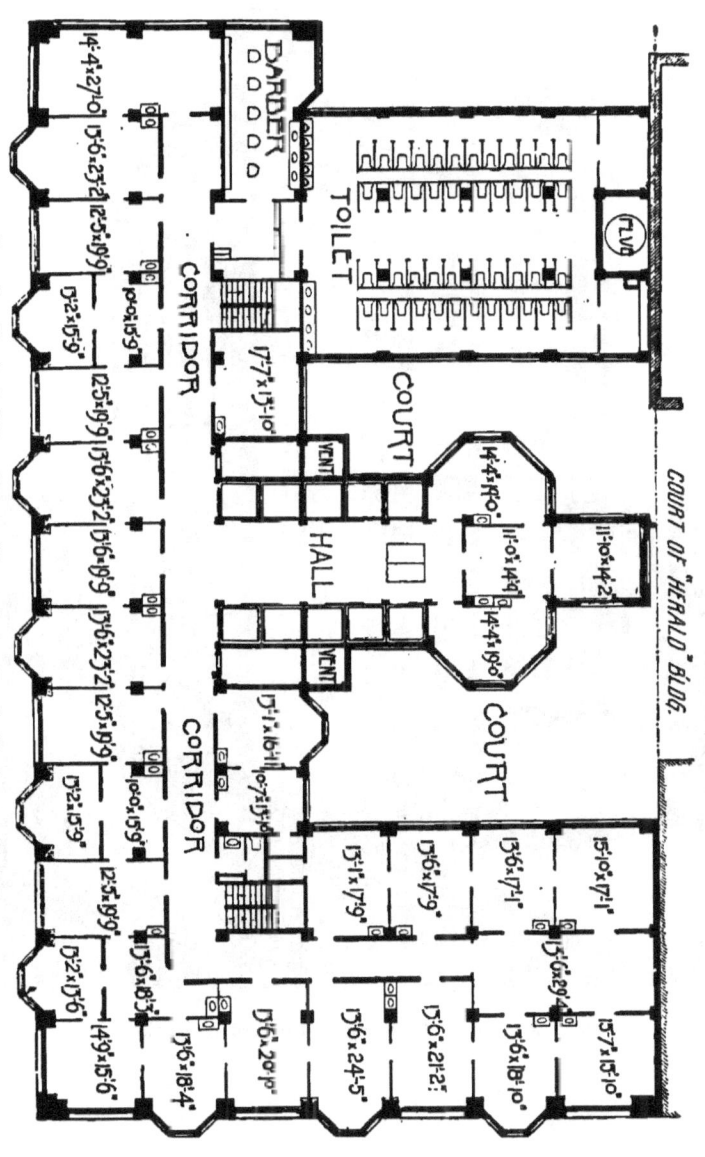

FIG. 6.—Typical Office Floor, Plan of the Stock Exchange Building.

FIG. 7.—The Marquette Building. Holabird & Roche, architects.

SKELETON CONSTRUCTION. 31

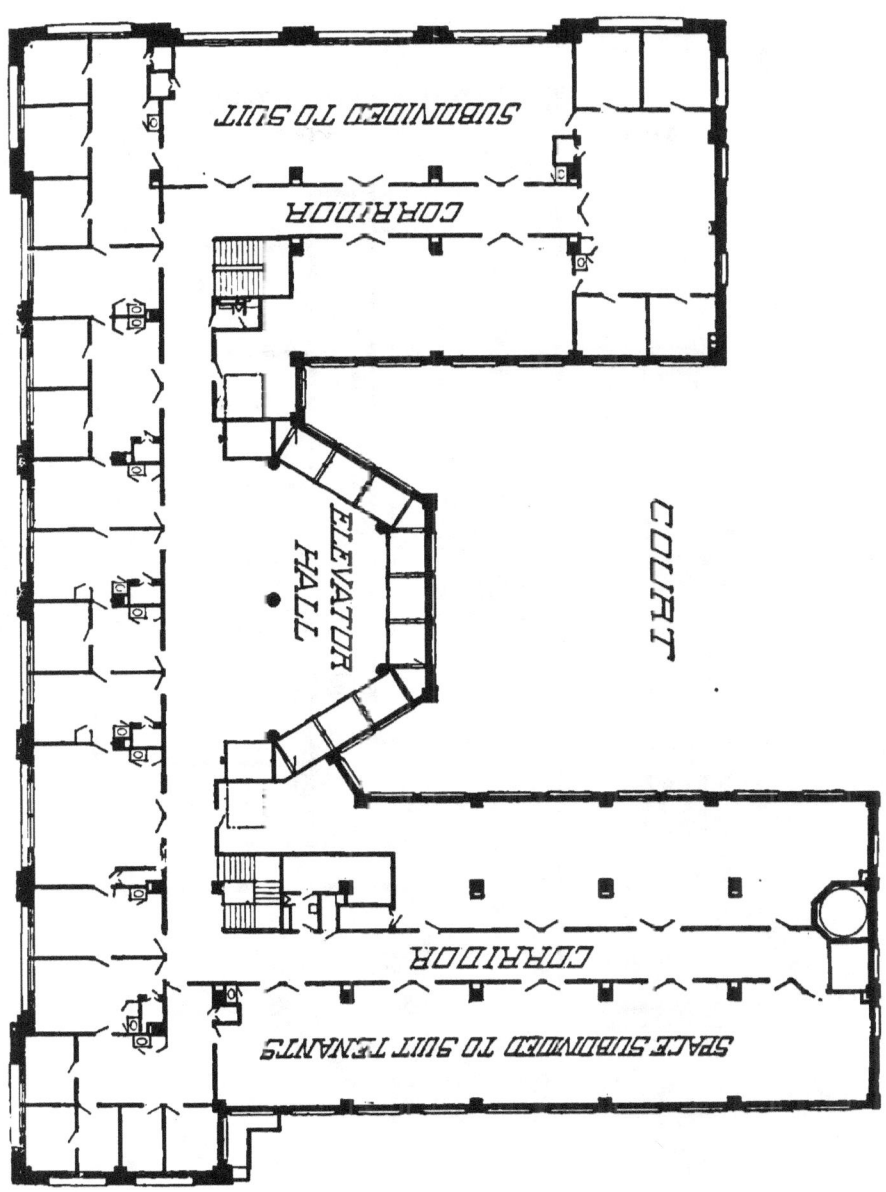

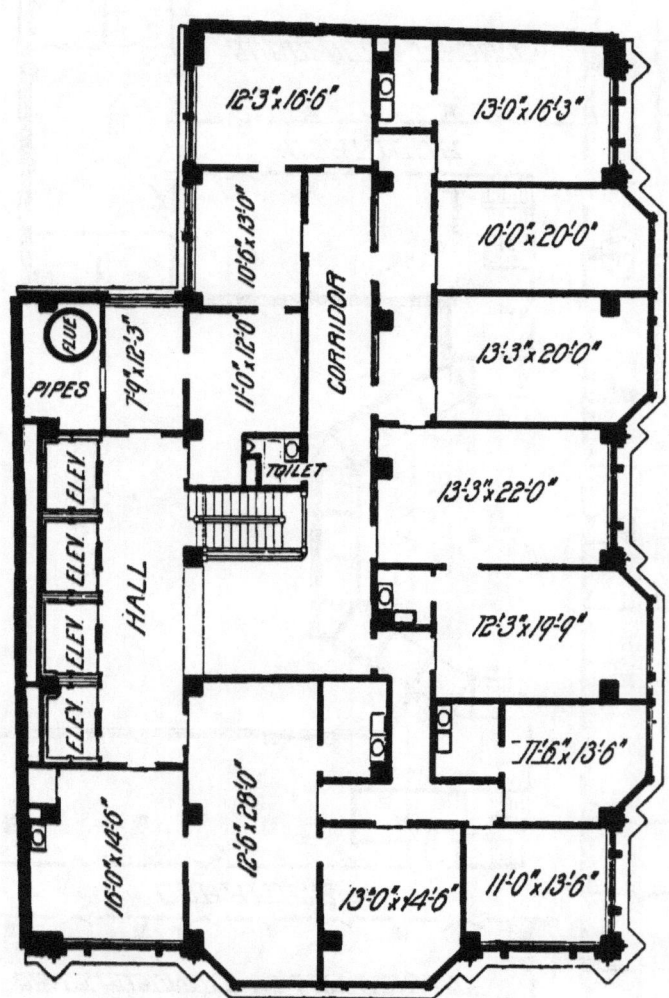

Fig. 9.—Typical Office Floor Plan of the Reliance Building.

supplied to support the elevator sheaves, and water-tanks located to supply the hydraulic cylinders.

If the basement, as in Fig. 4, lies below the sewer level, and it is to be occupied by stores, cafés, or by the boiler- and engine-rooms, an ejector pit will be necessary to raise the sewage to the proper level. Pumps for water-supply, dynamos for electric light, boilers and steam plant for power and heating—all must be definitely determined and carefully weighed in their relation to the character of the building, and as affecting the design of foundations and all structural details.

The following data may be of interest as descriptive of some of the mechanical furnishings of one of Chicago's most celebrated office buildings, the Masonic Temple, shown in Fig. 10.

The entire drainage is carried through the building by means of a system of vertical risers, about one half of which connect directly with the street mains, through piping suspended from the basement ceiling. The remainder of the risers, and all drainage from the boiler-room and basement space, are connected by a system of underground piping, with two 50-gallon Shone ejectors, placed in a pit in the basement, from which the sewage is forced to the street sewer. This was necessary in order to keep the basement stores, cafés, etc., free from exposed pipes. All vertical pipes in the building, both for water-supply and drainage, were carried in fire-proof pipe-spaces especially provided. The water-supply is pumped from the city mains, by pumps located in the basement, to storage tanks on the twentieth floor, with a combined capacity of 7000 gallons. On the twentieth floor also are four compression elevator tanks of 18,500 gallons capacity total. For elevator and water-supply service seven pumps are required, having a total capacity of from 2000 to 3800 gallons per minute.

Fig. 10.—The Masonic Temple. Burnham & Root, architects.

Each office and store has a private wash-basin, with general toilet-rooms and barber-shop on the nineteenth floor. The main toilet-room contains 64 closets, besides additional rooms on the third and twelfth floors and in the basement, with from 8 to 18 closets each.

FIG. 11.—The New York Life Insurance Building. Jenney & Mundie, architects.

Forty thousand square feet of radiation surface are required, all in direct radiation. The steam is supplied on the "overhead" system through 16-in. mains running directly to the attic, thence around the exterior walls and down. Six dynamos supply 7000 16-candle-power lamps. For the power and steam plant eight horizontal tubular boilers are used, with a total of 1000 horse-power.

There are several features in the Masonic Temple design worthy of especial note. Several of the upper floors

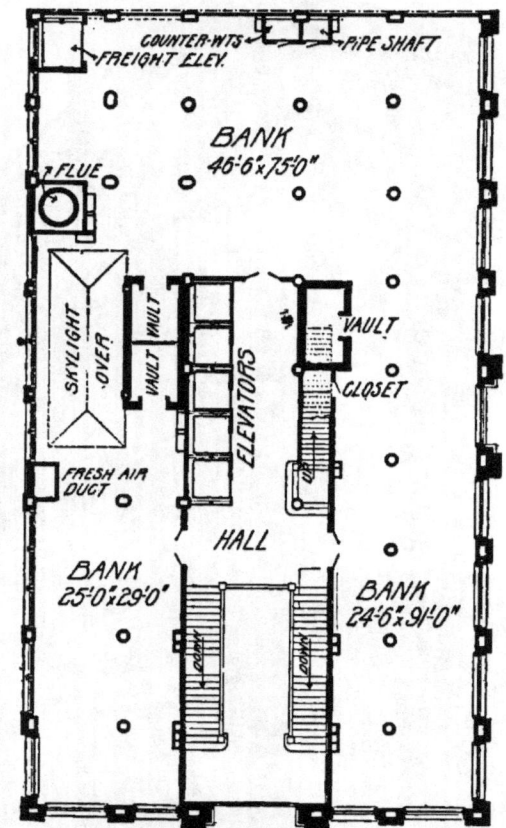

FIG. 12.—Banking Floor, New York Life Insurance Building.

are devoted to Masonic purposes, and the large assembly-, drill- and banquet-rooms were kept free from columns by spanning the areas with lattice girders, on which rest the arched ceiling and roof trusses. The interior court also possesses special features in the galleries provided at each story for the lower ten floors. This plan was intended to attract small storekeepers and the like as occupants of the adjoining stores or offices, thus concentrating many trades-

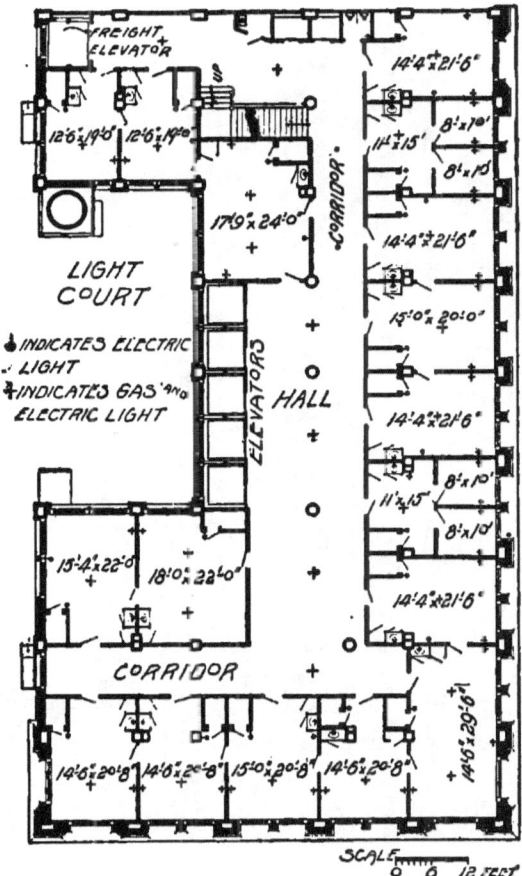

FIG. 13.—Typical Office Floor Plan, New York Life Insurance Building.

men under one roof. The scheme has not proved a success.

The roof of the Masonic Temple is covered by an enclosure of glass, serving as a summer-garden and place of observation.

NEW YORK LIFE INSURANCE BUILDING.

A perspective of this building, designed by Jenney & Mundie, architects, is shown in Fig. 11. The lower three floors are built of granite, with brick and terra-cotta above. The plan of the first floor, devoted to banking purposes, is shown in Fig. 12, while the typical office plan is shown in Fig. 13.

FIG. 14.—The Fort Dearborn Building. Jenney & Mundie, architects.

FORT DEARBORN BUILDING.

This building, shown in Fig. 14, is but just completed. It was designed by Jenney & Mundie, architects, and a number of the details used in its construction will be given later. The typical office floor plan is given in Fig. 15.

A floor plan of the Champlain Building, Holabird & Roche, architects, is shown in Fig. 16.

Fig. 17 gives a perspective of the Old Colony Building, by the same architects.

These examples of floor plans will serve to show the general arrangement of offices, halls, and entrances in buildings very recently erected, and the conditions which determined the general features of construction will be apparent, in so far as the plan may affect the locations of the columns, etc. "The framing plans" must now be worked out, one for each floor, showing the location of all piers, columns, girders, beams, etc., in their proper positions, with all the necessary dimensions and sizes.

Fig. 18 shows a framing plan of the third floor of the Fort Dearborn Building.

Fig. 19 is a framing plan for the sixth, seventh, and eighth floors of the Reliance Building.

The increased use of structural steel, as indicated in these framing plans, has found the architects, to a great extent, unprepared to solve in detail many of the problems imposed on them. They have been forced, in work of any magnitude, to turn the details, if not the entire constructional scheme, into the hands of the engineer, either as an employé or co-partner. The ignorance which the average architect displays in connection with structural iron details is proverbial, and contractors for steelwork especially have long indulged in considerable sarcasm at the expense of the architect and his plans. When, however, this work is intrusted to the engineer, it becomes a question as to how far the actual work of detailing needs to be carried, after the computations and general framing plans are made.

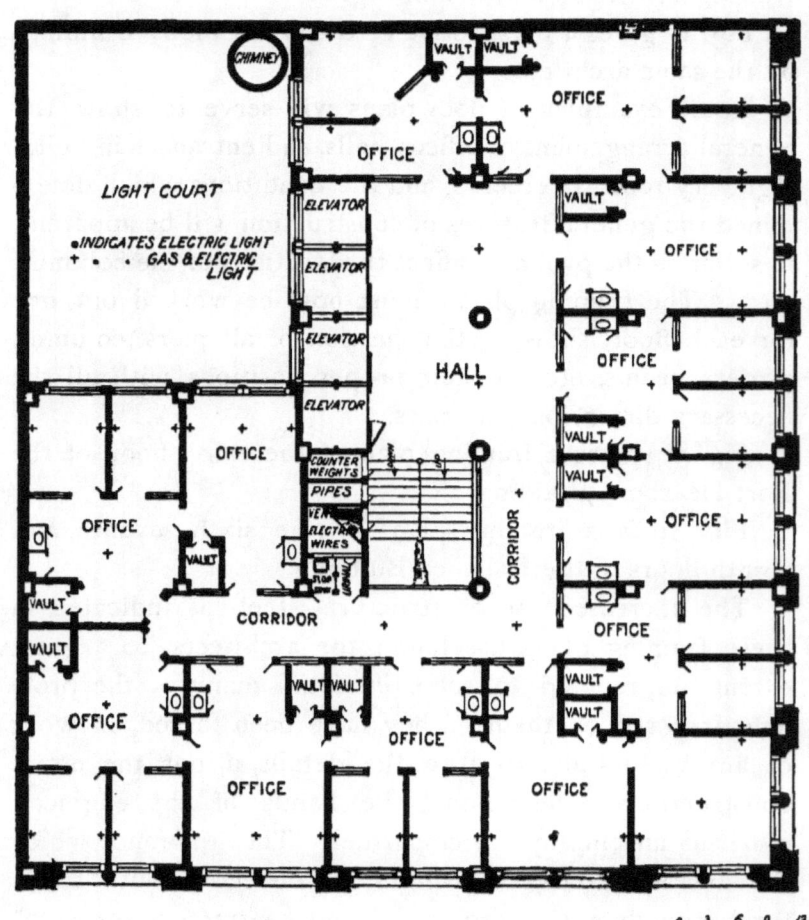

FIG. 15.—Typical Office Floor Plan, Fort Dearborn Building.

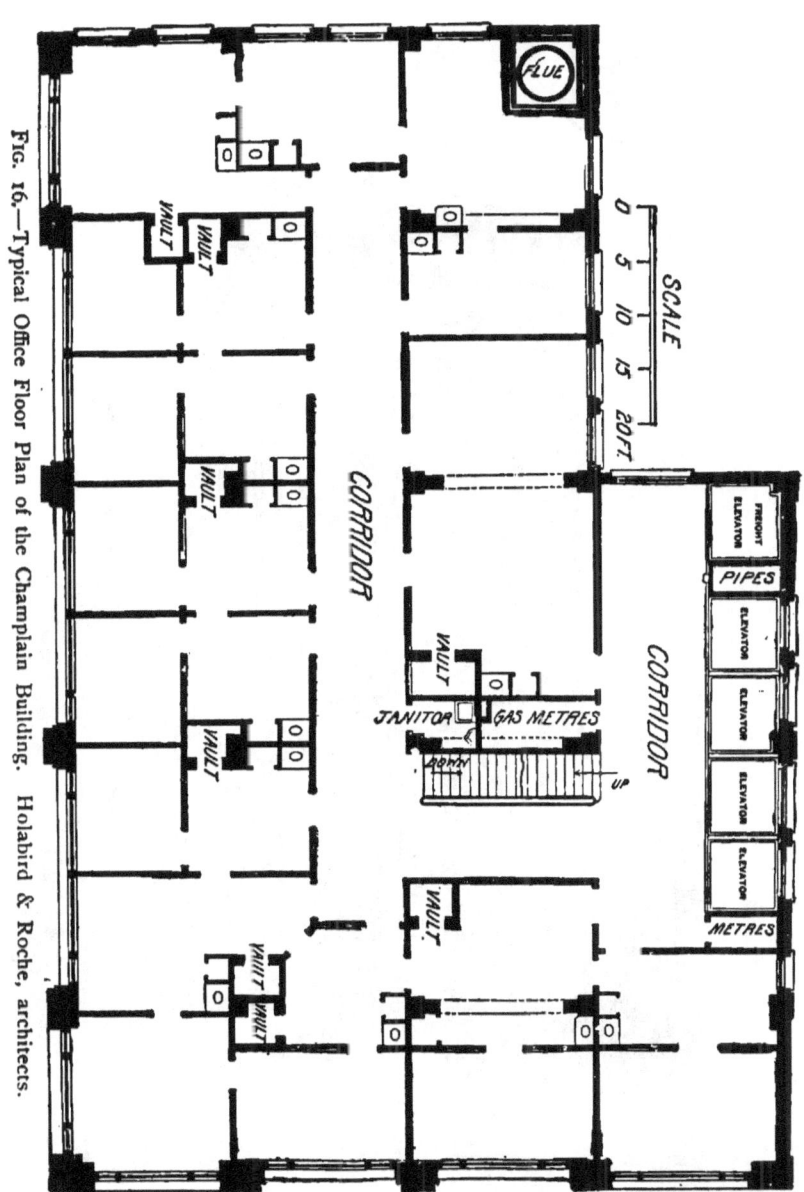

Fig. 16.—Typical Office Floor Plan of the Champlain Building. Holabird & Roche, architects.

FIG. 17.—The Old Colony Building. Holabird & Roche, architects.

SKELETON CONSTRUCTION. 43

It is comparatively seldom that complete detail plans for the steelwork of a building are made by the architect. Still less frequent are the cases where such detail plans could be used as actual shop drawings by the contractor,

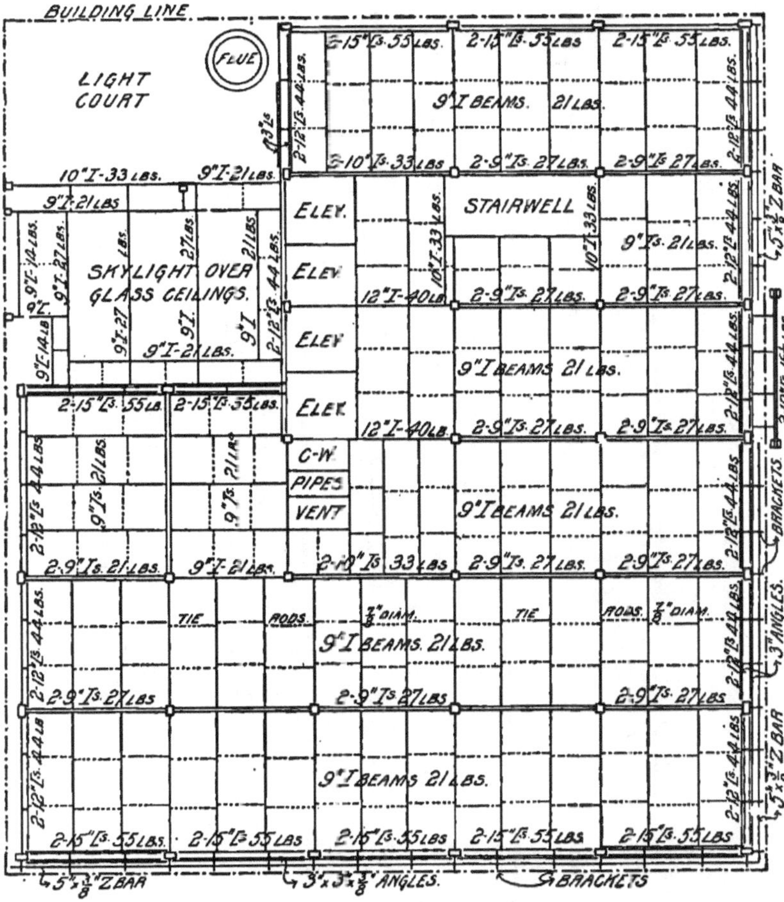

FIG. 18.—Typical Framing Plan of the Fort Dearborn Building.

as in nearly every case the manufacturer much prefers to make his own shop drawings, to conform to the usage of his own plant. The architect has generally been content to specify the sizes and weights of the material to be used, leaving the details to be worked out by the contractor with the approval of the architect.

The trained engineer, however, is not usually satisfied

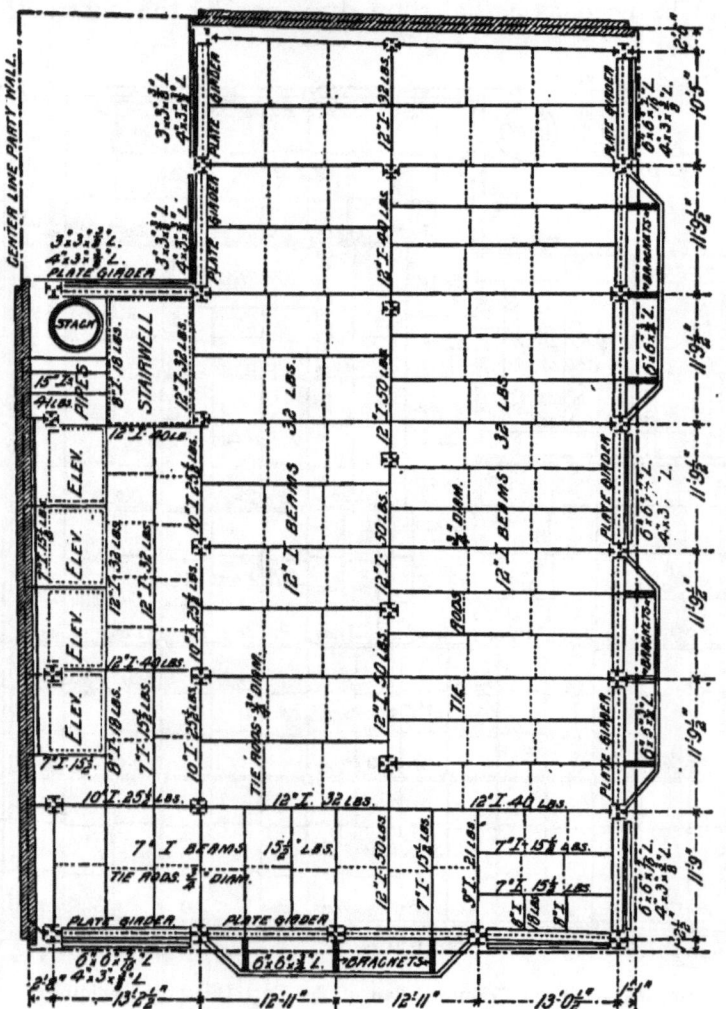

FIG. 19.—Typical Framing Plan of the Reliance Building.

with such license on the part of the contractor, and the best classes of work are made in accordance with definite details furnished by the engineer, after a careful consideration of the conditions to be fulfilled. This does not mean that complete shop drawings are made, but rather such connections and special points in the design as need particular attention. The balance of the detailing may be made to suit the contractor, with the approval of the engineer, in conformity with the sizes of material marked on the plan, and the carefully drawn specifications.

The idea of allowing the manufacturer to prepare complete details after his own general scheme, and following specifications only, is not consistent with best results, in the judgment of the writer, though such an arrangement has often been advocated. It is true that it has been a very common practice with bridge engineers to furnish the moving-load diagram, and allow the bidders to design the structure as they saw fit, so long as it fulfilled all requirements of the specifications. This has probably been one reason for the high degree of excellence shown in the work of the better bridge companies, as each bidder endeavors to use his material to the best possible advantage. Such a practice, however, in building work will require a very careful supervision of the work by the engineer, and as the various contractors will use those shapes most in favor, or of least cost, at their particular works, the calculations, connections, details, etc., must all be gone over and thoroughly checked, that all conditions may be satisfactory. A careful checking is necessary in any case, but where such complete freedom is accorded the bidder, it will rarely be that he is able to grasp the general ensemble in such a manner as to make satisfactory details in the required time. Again, only the most responsible and experienced firms could be intrusted with such a task.

Carefully drawn specifications, complete and accurate framing plans, sufficient spandrel sections and any special details, with all sizes and dimensions of material, will insure rapid and satisfactory work on the part of the iron contractor. The shop drawings may then be examined, and stamped with the approval of the engineer as received.

ERECTION.

In skeleton construction, the erection of the framework progresses very rapidly after the material is once delivered on the ground. All punching and riveting of the members is done at the shop, leaving only the assembling and field-riveting to be done on the ground, besides the adjustment of the laterals. Field-riveting has entirely superseded the use of bolts in the best class of work. Bolt connections were tried, but were soon discarded on account of the cracks which developed in the plastered ceilings, radiating from the column connections with the floor system. This was due to the play of the bolts in the holes.

Steam cranes built expressly for the purpose have been used in some cases in Chicago. They were operated on tracks which were quickly laid over the floor system, and these cranes would pull themselves up an incline, from story to story, as fast as erected. The crane boom and engine platform revolved on a pivot, so that the members required very little handling. The old-fashioned derricks or gin-poles are, however, generally used, some contractors preferring the short gin-pole, erecting one story at a time, while others use a large boom derrick, setting several stories in place before shifting the derrick. The erection of ironwork costs from $6 to $8 per ton.

Two stories can generally be erected in six days of ten hours each. In the Unity Building of seventeen stories the metal-work, from the basement columns to the finished roof, was accomplished in nine weeks.

The following data will give a better idea of the rapidity of building operations in Chicago as shown in the erection of the New York Life Building:

July 17. Old building torn down to grade.
July 31. Laid out new footings.
August 17. Started setting basement columns.
August 31. Started laying granite.
September 5. Started setting tile arches.
September 18. Started laying terra-cotta facing.
September 29. All steel set.
November 9. Tile floors all set.
November 11. Terra-cotta all set.
November 12. Started plaster.
December 2. Steam plant completed—turned steam on in building.

Of the 671 individual columns in this building but a single one required "shimming." A thin steel wedged plate was used, forged to fit. The columns were tested for alignment at frequent intervals. An average of twenty-five working hours was required to set the steelwork for a complete story.

The skeleton or "veneer" type of construction possesses great advantages in economy of time required for erection, as work can be pushed on the walls at different stories at one and the same time. Thus on the Manhattan Building, Chicago, the main cornice of terra-cotta was completed before the wall was built up beneath it. On the Unity Building the granite base-wall was being built at the first and second stories, the pressed-brick face was being placed at the twelfth-floor level, while the hollow-tile arches were being set for the fifteenth floor,—all at the same time.

Several gangs of men may frequently be seen at different levels on a single front of a building, and laying pressed-brick by electric light was even tried on the Ashland Block, Chicago, in an endeavor to complete the building by May 1: and the intention was to make up for this extra expense of night-work by time gained through leases signed earlier than would otherwise have been possible.

48 ARCHITECTURAL ENGINEERING.

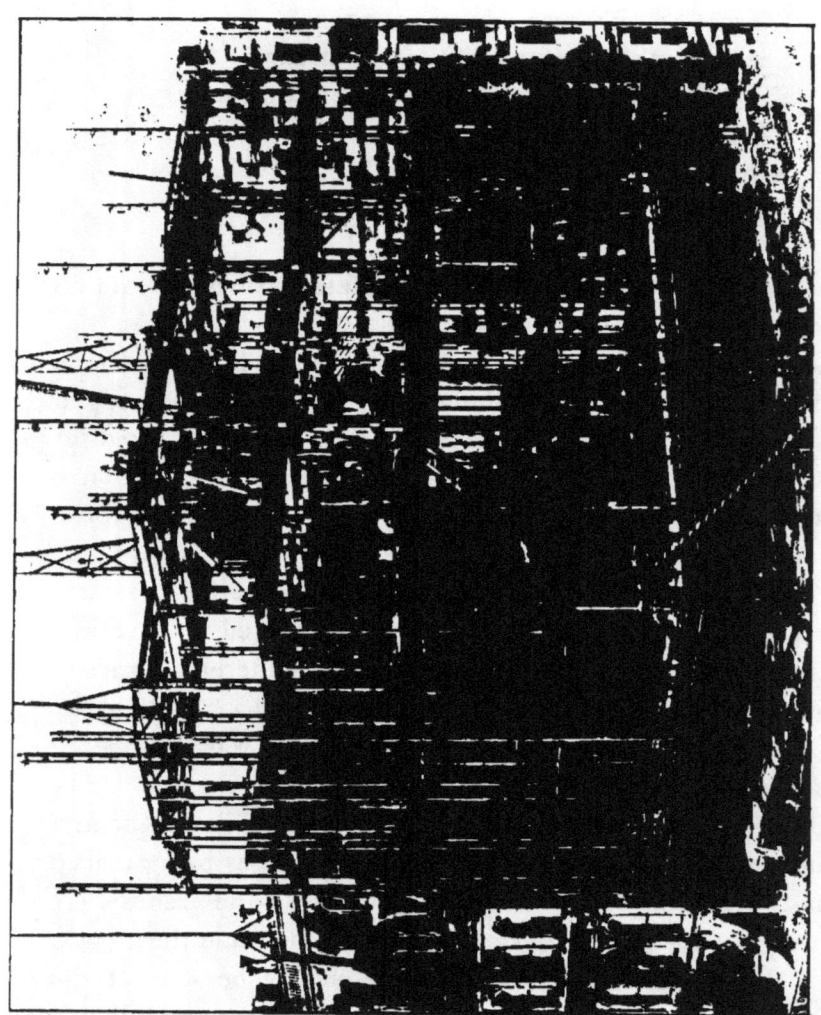

FIG. 20.—The Reliance Building during Construction, July 16, 1894.

FIG. 21.—The Reliance Building during Construction, August 1, 1894.

Figs. 20 and 21 show the Reliance Building during construction.

PERMANENCY OF SKELETON CONSTRUCTION.

Aside from the question of fire resistance, considerable discussion has arisen of late concerning the permanency of skeleton construction. This controversy between friends and indifferent observers of skeleton methods has been aggravated by the reluctance of the supervising architect of the Treasury seriously to consider such construction as worthy the dignity and solidity of government edifices—notably in the proposed new Post-Office building for Chicago. While the architectural pros and cons of terracotta and steel, or concrete and steel, versus solid masonry construction may not here be gone into, the engineering side of this matter becomes one of great importance. Serious as it is, it must still be admitted as depending largely on personal views, for the want of reliable data under present conditions. Many architects are not slow to pronounce judgment against such practice, while others warmly champion the cause of steel in combination with tile, concrete, or cement. The divergence of present opinion was well shown in a recent discussion before the American Institute of Architects on this very subject, where examples of the deterioration of iron or steel under peculiar conditions were emphatically offset by instances of remarkable preservation under other peculiar conditions. The point would then seem to be to *define* these conditions. Prominent Chicago architects and engineers have said that experience seems to show that if *no* lime mortar is used the corrosion of the metal will not amount to enough to be of any danger, while others point to the well-known *preservative* qualities of lime, and urge its exclusive use in connection with iron or steel. Our knowledge of wrought iron or steel, therefore, under definite variations of heat and moisture, and in association with limes, cements, and concrete, as found in present practice, must continue to be unsatisfactory until defined by more accurate data. Chicago engineers and builders show their daily faith in such combinations of material, and this type of construction is rapidly becoming more and more general in the United States.

The effects of lime, whether as one of the ingredients of mortar or limestone, as a corrosive factor in connection with ironwork seem to depend very largely upon the peculiar conditions of each particular case. Examples are recorded of anchorage cables in American suspension bridges which were found, on disclosure after some years, to be partly eaten away where the strands had come into permanent contact with the limestone masonry. The presence of water was possibly accountable for this corrosive action; but it becomes a very difficult matter to construct masonry which will allow of no permeation of moisture, especially in walls, piers, or foundations, as found in building practice. Dry air and pure water produce but slight oxidizing effects on iron or steel; " but when the former becomes moist, and the latter impure or acidulated, oxidation of the material is speedily set up, and when once commenced, unless the process is arrested, its ultimate destruction becomes a simple question of time." The use of lime mortar would, therefore, seem limited to localities where no fear of moisture may be anticipated; for any dampness in combination with the lime, must soon show its effects on the metal-work.

Considering the parts of a skeleton structure which are exposed to the weather, or liable to the presence of moisture, we have: all exterior walls, piers, etc., and the basement members, including foundations. From the foregoing it would seem that lime mortar should not be used in any of these positions. The foundations and basement walls, columns, etc., are either surrounded by constant moisture, or by wet clay or earth itself, while the exterior walls and supporting steelwork are subjected to the climatic changes, frost, rain, and penetrating dampness, which must sooner or later pierce the terra-cotta and brick envelope, and so reach the metal-work. For such positions cement mortar should undoubtedly be used; it seems a most perfect conservator of metal-work, and instances are recorded of iron found in perfect condition after a 400-years' entombment in cement concrete below water. Links of anchorages in American suspension bridges have been taken up after many years in a perfect state of preservation where embedded in cement. A further recommendation of the use of cement lies in the fact that the thermic expansion of Portland cement is practically the same as that of iron—a fact which insures perfect cohesion under any changes of temperature.

The interior members of the framework do not need as careful consideration, being maintained at a more uniform temperature, and protected from the exterior dampness. Interior columns, the floor system, and wind bracing would, therefore, seem safe in connection with lime mortar, but it is questionable whether the best work should not call for cement mortar and even cement plaster throughout. Cement has rapidly cheapened of late years, and cement plasters are largely being used on account of their better fire-resisting qualities.

It has been suggested to rely entirely on the preserving qualities of cement rather than on a proper painting of the metal-work. Prof. Bauschinger states that his experiments show a cohesion between iron and concrete after hardening of from 570 to 640 lbs. per square inch. This is even more than the tensile strength of the best concrete, but in building work a perfect union between the cement mortar and metal-work can never be attained at all points, and a thorough coating of paint must largely be relied upon.

All constructive ironwork should, therefore, be well coated with either lampblack mixed with oil, or red lead and linseed-oil. The very best of materials should be employed. The oxide of iron or mineral paint which has generally been specified for all painting of the metal-work has been found to separate from the steel, and form an oxidation of the metal behind the paint. A mixture of red lead and linseed-oil is now considered as the best protective coating for iron or steel. A careful inspection of all painting, both at the shop and in the field, should be rigidly enforced.

The following are the requirements of the New York building law in regard to the protection of iron or steel work against rust, etc:

"All ironwork and steelwork used in any building shall be of the best material and made in the best manner, and properly painted with oxide of iron and linseed-oil paint before being placed in position, or coated with some other equally good preparation or suitably treated for preservation against rust."

The Chicago ordinance makes no mention of paint or coatings to prevent rust in the metal framework except as specified for fire-proofing purposes as follows; "In all cases the brick or hollow tile shall be bedded in mortar close up to the iron or steel members, and all joints shall be made full and solid."

The Boston law requires a protection from heat only, by means of brick, terra-cotta, or by three fourths of an inch of plastering.

The requirements for metal-work in foundations are given in Chapter XII.

CHAPTER IV.

FLOORS AND FLOOR FRAMING.

THERE is scarce a subject or detail in the present field of architectural engineering that has provoked such widespread attempts at improvement and perfection as the question of fire-proof floor systems. The present day is especially prolific in new patents and systems, all claiming a complete revolution in existing methods, until both architect and engineer alike are well-nigh bewildered in their endeavors to keep track of the novelties that are continually being presented as the " cheapest and best " solution of a much-discussed problem.

A proper solution cannot be realized by either architect or engineer working independently of each other, and perfection in present attempts must result from legitimate criticism on the part of the architect as to the adaptability of the material to exterior form, as well as from the application of the laws of statics as demanded by the engineer.

Before investigating present methods and future probabilities it will be profitable to examine earlier systems, with their weak points and causes of failure.

The oldest so-called fire-proof arches consisted of I beams, placed about 5 feet centres, with 4-inch brick arches turned between, then levelled up with concrete containing the nailing-strips for the wooden flooring. Corrugated iron, sprung from flange to flange, was also used in place of the brickwork, and this latter type may still be seen in some of the more substantial buildings of that epoch, which have survived to the present time. This construction was decidedly faulty, however, not alone in the weakness of the arch itself under the action of fire, but in the fact that the lower flanges of the supporting I beams, and the entire cast columns then in use, were left exposed to view, and, what was much more serious, to the possibility of contact with fire.

This unsatisfactory and weighty construction, shown in Figs. 22 and 23, gave way, as has been said, to the superior

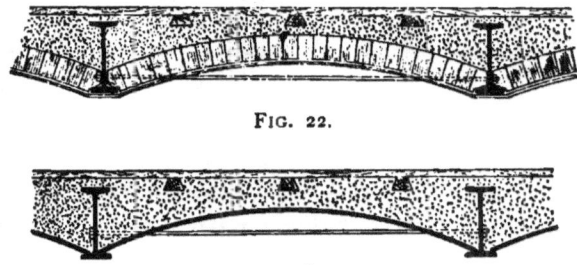

FIG. 22.

FIG. 23.

advantages of terra-cotta or tile—superior in fire-resisting qualities, as well as in greater lightness.

Hollow tile is made from fire-clay, moulded by dies into the various hollow forms required for commercial use. The clay is subjected during its manufacture to a high pressure while in a moist or damp state, which accounts for its great strength, and after drying is burned, like terra-cotta, in a kiln.

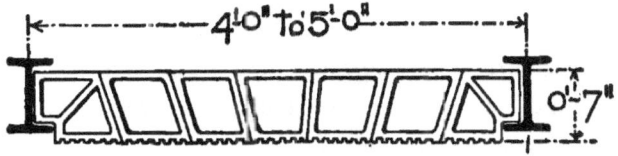

FIG. 24.

The clay used in the manufacture of fire-proofing material must be of a refractory nature—as plastic fire-clay, semi-fire-clay, or fire-clay mixed with plastic clay or shale. But few clays have been found that are practicable of manufacture into a floor of the required strength, and reliability against fire.

TILE ARCHES.

The earlier forms of tile arches were made as in Fig. 24, which shows the arch used in the Equitable Building in Chicago (1872), and Fig. 25, which shows tile arch in the

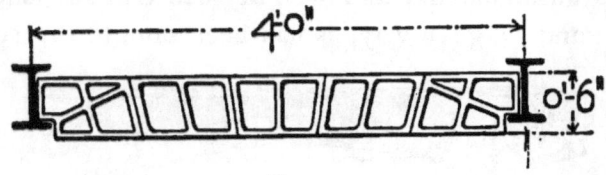

FIG. 25.

Montauk Building, Chicago (1881). The latter may be said to have been the first building of modern design in Chicago. The arches were 6 inches deep, with a span of 3 to 4 feet. But as these forms still left the lower flanges of the I beams unprotected, they were soon superseded by the type shown

FIG. 26.

in Fig. 26. This arch was used in the Home Insurance Building, Chicago (1884), the tile being 9 inches deep and 6 foot span. This was the first instance in which the beam soffits were protected against fire by anything more than plaster; and as many of the features in this arch are essentially the same as in the types of tile arches as found in present practice, a brief description will here be in place.

The pieces form radial joints, as in any segmental arch, or are key-shaped with a centre "key." The arches are set on "centres" of plank, hung from the beams by hook-bolts, and these centres should remain in place at least twenty-four hours after the arches are set. The "skew-backs," or butment pieces of the arch, take the shape of the I beam against which they bear, setting firmly and squarely on the beam flanges. Different sized skew-backs are at hand for use with different sized beams, as arches are often sprung between beams of different depths. The soffit of the tile arch extends about one inch below the bottoms of the beams, and the skew-back pieces are made in such a manner that a piece of fire-proofing tile may be slipped in and supported directly underneath the beam flange, to complete the fire-proofing, as shown in Fig. 26. A coat of plaster or cement is then given the whole surface, after which it is ready for such decorative treatment as may be desired.

A concrete filling is placed over the arch, to distribute the load from block to block, and to receive and embed the wooden nailing-strips which take the finished flooring. The metal beams are thus entirely surrounded by fire-clay, concrete, and cement.

The depth of the tile arch depends upon the span, and the load to be carried. The maximum spans of the various

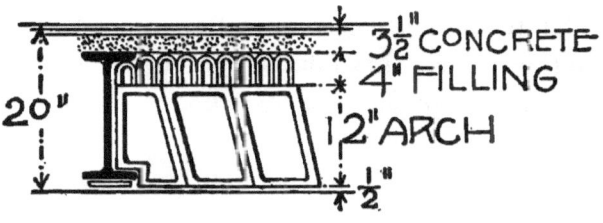

FIG. 27.

depths are generally furnished by the manufacturer of the type in question, but such data should be fully established by adequate *tests*, as will be pointed out later. Slight variations in the span from centre to centre of beams are made by using "half intermediate" tile, and different-sized keys. The tile blocks are laid with lime mortar or cement joints, and in no case should the joint exceed ½ inch in thickness.

In many cases, where the panel length required beams of a considerably greater depth than the tile arch itself, tile filling-blocks were used, as being lighter than the ordinary concrete filling—as shown in Fig. 27, taken from the Woman's Temple, Chicago. Special shapes for skewbacks, panelled beams, etc., made in this character of tile, are shown in Figs. 28 and 29.

Fig. 28.

Fig. 29.

The best semi-porous tile used in these types was made from clay found at Chaska, Minn., at Brazil, Ind., and in parts of eastern New Jersey.

In the foregoing examples of arches, known generally as the "Pioneer" arches (because made by the Pioneer Fire-proofing Company of Chicago), the voids in the tile blocks ran parallel to the supporting beams, and hence the principal or side webs of the individual tile blocks also ran parallel to the beams, or at right angles to the line of thrust in the arch. This limited the effective arch area to the top and bottom flanges, involving a serious waste of material.

To remedy this defect a new arch was patented a few years ago, known as the "Lee" arch, in which the voids ran parallel to the line of thrust, or at right angles to the supporting beams. One of these arches is shown in Fig. 30, and it will be seen that the effective area now comprises the vertical webs, as well as the horizontal ribs; in other words, all of the material performs useful work *as an arch*. A further improvement was attempted by the use of a porous terra-cotta, made from a fire-clay which, before it is burned, is mixed with sawdust and finely cut straw. These ingredients are consumed during the firing, leaving the material in a very porous condition, and thus greatly reducing the dead

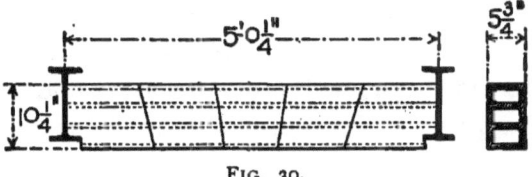

FIG. 30.

weight of the arch itself. A comparison of the weights of the old Pioneer and the newer Lee arch may be made as follows (weight given is per square foot):

	Pioneer.	Lee.
9" arch	33 lbs.	25 lbs.
10" "	37 "	30 "
12" "	40 "	35 "
15" "		40 "

Another step of progress lay in the skew-back or butment pieces, which gave a better bearing against the beam webs by means of intermediate cross-ribs, as well as by the top and bottom flanges.

Some very interesting and valuable tests of fire-proof floor arches built after the Pioneer and Lee methods were published in No. 796 of the *American Architect and Building News*—undoubtedly forming one of the most satisfactory and extensive series of public tests yet attempted on such construction. The trials were made in Denver, Col., 1892, for the Denver Equitable Building Company, under the supervision of a board of architects. The arches were sprung from beams placed 5 ft. centres, as shown in Fig. 30, and the conditions included static loading, a drop test, a fire and water test, and a continuous fire test.

In the test for static loads the Lee arch deflected gradually under the increased weights to .065 of a foot, sustaining a final load of 15,145 lbs. for two hours. The Pioneer arch gave way suddenly at the haunches under a load of 5,429 lbs.

In the drop test a piece of wood 12″ × 12″ × 4′ was let fall from a height of 6′ 0″. The Pioneer arch was shattered at the first blow, while the Lee arch, under the same test, stood up to the eleventh drop, the former blows shattering but parts of the arch.

In the fire and water tests, three applications of water combined with fire destroyed the Pioneer arch, while the Lee arch received eleven applications of water, and at the end of twenty-three hours remained practically uninjured, requiring eleven blows from the ram to break it.

In the continuous fire test the fire was maintained continuously beneath a Lee arch for twenty-four hours, and the arch then supported a load of bricks of 12,500 lbs. on a space 3′ 0″ wide in the central portion of the arch.

Considering the static loads, the results may be better judged as follows:

	Pioneer.	Lee.
	lbs.	lbs.
Breaking-load per square foot of 9 sq. ft. loaded area.	603	1670
Reduced to equally distributed load, 3′ 0″ × 5′ 0″	360	1008
Assumed load per square foot, as occurring in practice	150	150
Coefficient of safety	2.4	6.7

This certainly shows a great step of advancement for the Lee arch, but, assuming a factor of safety of 8, as recommended by Rankine, and a total load of 165 lbs. per square foot (85 lbs. dead + 80 lbs. live), a *uniform* breaking-load of 1320 lbs. per square foot is needed before the tile arch can be considered fully acceptable.

Tests of the 12″ blocks of the Empire Fire-proofing Company might also be mentioned, made in 1891 by the city engineer of Richmond, Va. A variation in the breaking-load was recorded of from 554 to 1057 lbs. per square foot. But it must be remembered that too much importance must not be placed on these maximum figures. The *average* breaking-loads of such tests must be considered a fair figure at which to judge the general run of arches as placed in actual use by these companies; and in this light the room and actual necessity for still further improvement becomes self-evident.

A still later patent known as "Johnson's patent flat arch," (see Fig. 31), is the one used most extensively in the

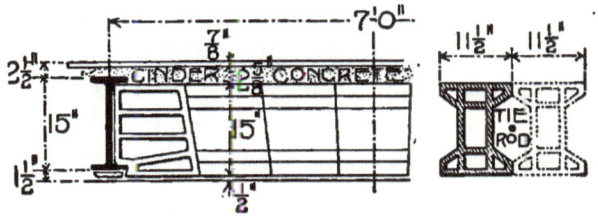

FIG. 31.

buildings of late erection. It is made of hard terra-cotta with thinner webs than were formerly employed, and is of the "end construction," thus utilizing all of the material as in the Lee arch. This type seemed to meet with much favor at first, and it was used in quite a number of Chicago's best buildings, but experience would seem to point to the porous tile as being far more satisfactory in its fire-resisting qualities than the hard tile. A test by fire and water of a wall of hard tile blocks occurred some time ago in the rear of the Schiller Theatre Building, Chicago. The combined action of heat and cold water caused the blocks to crack to such an extent that they soon fell from the metal uprights in considerable areas.

Soft tile or porous terra-cotta has been specified for all fire-proofing work in the latest buildings designed by Mr. W. L. B. Jenney, notably the New York Life Insurance and the Fort Dearborn buildings.

Tie-rods are necessary in all these forms of arches, to take up the horizontal thrusts without dependence on the adjoining arches. Such rods are generally ¾ inch diameter, and spaced from 5 to 7 feet apart. All tests of tile arches should require the tie-rods to be without initial strain; for if the rods be screwed up sufficiently to give an initial strain equal to the tensile strength of the tile or cementing material between the blocks, then is the tensile strength of the arch for the breaking-load reduced to 0, and the beam may be reloaded to the same amount.

Reference to the Appendix table giving the principal points of construction in the notable office buildings in Chicago will show that either the "Pioneer," "Lee," or "Johnson's" type of floor arch is used in nearly every case, although it must be admitted that such a general use of tile construction is far from being a guarantee of its perfection. Indeed, it is no exaggeration to say that there is scarcely a single material used in constructional work in regard to which we have as limited a knowledge of its general or specific properties of resistance as is found in terra-cotta or tilework; and yet, in the modern building the use of this style of floor has become so widely extended that terra-cotta or hollow tile has become one of the most ordinary materials of construction. Its functions are no less positive than those of the structural steelwork or masonry-work, forming, as it does, the supporting area for all dead and live loads coming on the floor system—crowds in halls, theatres, and other places of public gathering, as well as small safes, desks, and the many articles forming concentrated loads.

Any failure in the hollow tile would be apt to result in quite as great disaster or loss of life and limb as would proceed from any failure in the iron or steel skeleton. It is apparent, therefore, that the sustaining power of hollow-tile work should be absolutely definite, and that its use should be governed by well-defined tests, or rules based on such tests.

Some attempts on the part of the writer to secure reliable facts pertaining to tests on some of the newer tile floor arches in almost daily use, developed the fact that the fire-proofing companies had no data " in shape to be made public," although the very types of arches about which information was asked, had already been used in a number of prominent buildings. This leads to the opinion that architects and owners are too free in accepting the *alleged* results of tests, or in accepting some general style of arch because it has been used elsewhere. When iron and steel specifications require severe tests from the finished material, representing each blow or cast, it would hardly seem more unreasonable to require actual tests for the style of arch used before accepting such arches for any particular building. It is only by such repeated tests, and competition based on actual results in each instance, that the most economical designs for floors can be obtained, consistent with good engineering principles; and it is certain that such tests and open competitions would lead to better quality, if not to better forms and details. Both concentrated and uniform loads should be considered, as occurring in actual circumstances.

The most satisfactory set of public tests on tile arches, in which quality and not price has determined the award of the contract, were those made for the Equitable Life Insurance Building in Denver, Col., already referred to. A load of 1008 lbs. per square foot equally distributed was carried in this instance, and though even higher figures are claimed for later patent arches, the prime consideration of price in the usual letting of contracts will soon tempt the less reliable fire-proofing companies either to juggle their figures into deceiving records of tests, as is often done, or to furnish poorer and poorer material as competition increases and searching inquiry decreases.

Rankine advises the use of $\frac{1}{4}$ to $\frac{1}{3}$ the ultimate strength in metals, $\frac{1}{8}$ to $\frac{1}{10}$ in wood, and $\frac{1}{4}$ to $\frac{1}{8}$ in masonry. Considering hollow tile as coming under the head of the poorest class of masonry, an ultimate strength should therefore be required of eight times the allowable stress, if it is wished to procure *uniform* safety in a floor of steel beams and tile arches. Assuming an arch carrying a live load of 80 lbs. per square foot, a dead load of 85 lbs. per square foot (or 165 lbs. total), the manufacturer should be required to show by tests on the site that the type submitted is able safely to stand a load of 1320 lbs. per square foot, and this before being allowed to compete in the question of cost. The writer is aware of the objections of time and cost to such methods, but in this way only can the excellence be maintained, and assurance be provided that the floor arch is what it should be.

The unusual interest which is being displayed in the subject of fire-proof floors, of tile and other materials, is evidenced by the series of articles but lately begun in a periodical devoted to the interests of the clay products. This series of articles contemplates a complete record, as far as possible, of all tests on fire-proof arches of ordinary patterns, up to present date, with comments on the causes of failure and possibilities of improvement. Such work cannot fail to be productive of the most beneficial results.

The writer believes that the section devoted to the arch-like action in present tile floors is still too small. This is indicated by the sudden collapse of many arches at the haunches while under test. It would also seem, through past tests, that too much reliance has been placed in the use of strong cementing materials between the blocks, thus making the arch act as a monolithic piece. Hollow tile blocks, as used in present forms of arches, cannot be considered as a beam, even with the best of cement joints. They must still form a flat arch, whose line of resistance must be determined precisely the same as in any segmental arch. The fact that the arch blocks are of a uniform depth cannot in any way change the mechanical conditions under which the loads and supporting forces act.

Present types of tile arches do not admit of a proper calculation of their dimensions according to the loads for which they are designed; the horizontal bearing-ribs are still relied upon to help make up the required section, and the height of the section as well as the thickness of the tile webs, under different spans and loads, is left entirely to the option of the manufacturer. None of the building laws prescribe any conditions for the proper calculation of floor arches under varying spans and loads, except to define a minimum depth of arch blocks.

The depth of the tile arch should be nearly equal to the depth of the supporting I beams, in order to secure the most economical results, for this arrangement will be the cheapest in the cost of the floor per square foot, considering tile and concrete filling, and the lightest, considering the dead load.

An arch has been patented as shown in Fig. 32, but it is evident that the concrete or cinder filling at the haunches

FIG. 32.

will cause the arch to weigh more than if the tile blocks extended up to the tops of the beams; while the mere fact of the arch being made with a segmental top adds nothing to the strength.

CONCRETE ARCHES.

As has been stated before, the widespread interest displayed in the subject of fire-proof floors is indicated by the numerous types which have entered the field in competition with the hollow-tile flooring. It is certainly no difficult problem to design and construct a floor which will be of sufficient strength and of satisfactory fire-resisting properties out of fire-clay, cement, or concrete. But when the elements of minimum cost and minimum weight must be considered with maximum efficiency, the solution is not so apparent.

Up to the present day fire-proof floors have been enormously heavy, consisting largely of *dead* weight in the most literal sense of the word. Such weights add greatly to the cost of a building, and yet serve to little or no purpose in strengthening or stiffening the structure. Hence the endeavor to provide a substitute for the hollow-tile floor which shall yield an increase in unit strength, and thus decrease the dead weight and consequent cost.

A variety of combinations of iron or steel and concrete as applied to floorings has lately been employed, and would seem to possess features of great merit and of widespread application. Floors constructed of concrete and steel, with the latter thoroughly protected against corrosion, would certainly possess the great advantages of incombustibility and durability. It has long been claimed that the unequal rates of expansion and contraction of iron and concrete or cement under thermic changes would soon destroy such a combination, but experiments have been made which show that these rates are so nearly the same that they may properly be considered identical. The recent tests of such flooring at Trenton, N. J. (see *Engineering Record*, December 22, 1894), would also seem to point to the successful fire endurance of such combinations.

The weakest points against fire would appear to be the thin coating of cement plaster directly underneath the beam flanges, where a stream of cold water applied to the highly heated cement would probably cause it to crack off, and leave the metal-work exposed.

It is of great importance to ascertain these points by means of actual tests before final adoption, the same as in the case of tile arches; while the most judicious form of the metal-work, and the shape and character of the moulded concrete to develop the maximum resistance with the least weight, must also be determined by repeated tests.

The different character of the metal-work in combination with the concrete, presents three varieties: floors using curved I beams, those using steel straps, and those using wires.

1. *Curved I Beams.*—This system, shown in Figs. 33 and 34, is called the Melan system, from the inventor, J. Melan,

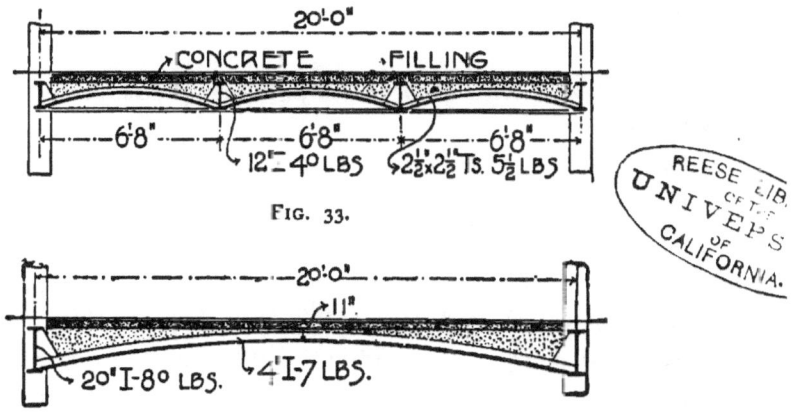

Fig. 33.

Fig. 34.

who has constructed many bridges of this type in Europe. It consists of bent I beams, spaced about 5 ft. centres, with concrete body or slabs between. A filling of cinders or other light material is then used to level up the surface and receive the nailing-strips, as in other floors. A great saving in dead weight and cost is claimed for this system, but it still possesses great disadvantages which, in the opinion of the writer, will seriously restrict its use.

The concrete must perform a twofold duty. It helps to take up the compression of the arch, and at the same time must act as a *beam* between the curved ribs. The fibres are then brought under maximum strain in two directions, and if we adhere to the usage of allowing no cement in tension, this combination becomes poor engineering practice.

In most cases where appearances are considered, a suspended ceiling will be necessary. Tenants and owners desire a ceiling of unbroken plane, for the sake of light as well as appearance. If such a suspended ceiling is to be of fire-proof construction, it will necessarily add materially to the weight and cost; or, if it is not of fire-proof material, a large amount of combustible material is added in a very dangerous position.

Exposed tie-rods are necessary, unless a suspended ceiling be used.

The workmanship must be of the most careful character, to insure the proper results from these concrete beams.

2. *Concrete and Steel Straps.*—Concrete floors in combination with steel straps have been used in the following buildings: Drexel Institute, American Philosophical Society Building, and Academy of Natural Sciences, in Philadelphia, and in the Alumni Building of Rensselaer Polytechnic Institute at Troy. This form of flooring, shown in Fig. 35, consists of I-beam girders spaced 8' 0" to 18' 0" centres as may be required, between which are hung steel straps at intervals of from 12" to 24",

with their ends bent or hooked over the top flanges of the girders. The straps curve downward, and midway in their length hang close to the ceiling-line. A concrete or

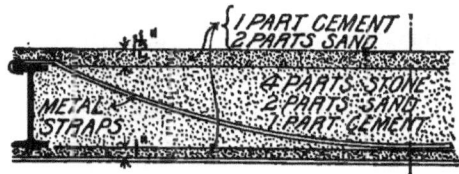

FIG. 35.

cement filling is used, embedding the straps and girders. If the beams are of considerable depth, the soffit of the arch may show the panelled beams, as in Fig. 36. The

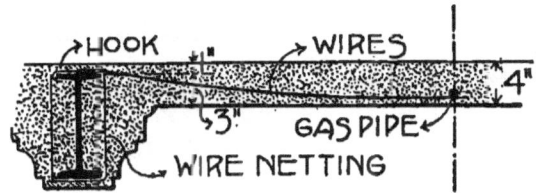

FIG. 36.

upper layer of cement may be laid in colored geometrical patterns, or, if a wood floor is used, this upper layer of cement is made but 1" in thickness, with nailing-strips embedded.

The following table gives data from two of the buildings before mentioned:

Building.	Span.	Straps.	Centre to Centre of Straps.	Thickness of Concrete.	Assumed Load.
American Philosophical Society Building.....	8'	$\frac{3}{4}'' \times \frac{3}{8}''$	24"	4"	80 lbs. live load.
	16'	$2\frac{1}{4}'' \times \frac{3}{8}''$	24"	7"	80 " " "
	16'	$2'' \times \frac{3}{8}''$	12"	8"	210 " total "
Academy Natural Sciences	18'	$2\frac{1}{2}'' \times \frac{3}{8}''$	18"	8"	100 " live "

3. *Concrete Floors with Twisted Wires or Rods* (see Fig. 36). —This method is very similar to the previous type, except that wires are used instead of straps. The wires are secured to the beams by means of hooks, 3" long, made of

¼" square iron. The wires are of twisted double strand, No. 12 gauge, with a length of gas-pipe laid on them at the centre of the span to give them a uniform sag. The filling consists of five parts by weight of plaster of Paris, and one part of wood shavings, mixed with sufficient water to bring the mass to the consistency of a thin paste. This filling is laid on a level centering, as in the previous type. The distance between the wires is varied, according to the load to be provided for.

Where a flat ceiling surface is desired, this type is modified, as shown in Fig. 37. The floor-plate is constructed on

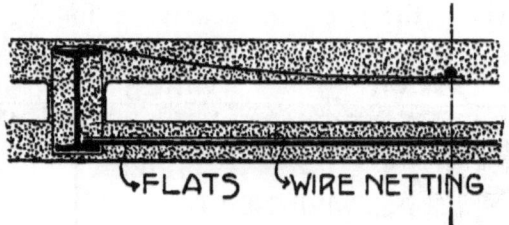

FIG. 37.

wires as before, while the ceiling-plate is made of the same composition, but with flat bars embedded therein, resting on the lower flanges of the I beams.

Tests of this flooring under static loads have been made as follows:

Distance between Beams, Centre to Centre.	Clear Span between Flanges.	Length of Section Tested.	Area Tested, Square Feet.	Total Load in Lbs.	Load per Sq. Ft. in Lbs.	Remarks.
4' 7"	4' 2"	1' 0"	4.166	5,630	1,351	Did not fail. Test made in building under contract.
5' 5"	5' 0"	0' 9¼"	3.958	7,600	1,920	Two cables on one side broke, others unbroken.
4' 6"	4' 0¼"	2' 6"	10.105	15,682	1,551	Failed by breaking of cables on one side.
4' 6"	4' 0¼"	5' 0¼"	20.38	29,314	1,438	Adjoining sections, being without load, lifted. No wires broken.

The fire and water tests also proved very satisfactory; indeed, plaster of Paris or gypsum has been used in Europe for many years as a fire-proof material.

The greatest objections to these arches lie in the discoloration of the plastered ceiling, due to the rusting of the wires, and the excessive amount of water retained for a long time by the sawdust. Galvanized wires should be used, and some material substituted for the sawdust.

Another form of concrete arch which has been used in California (see *Engineering Record*, March 24, 1894) depends

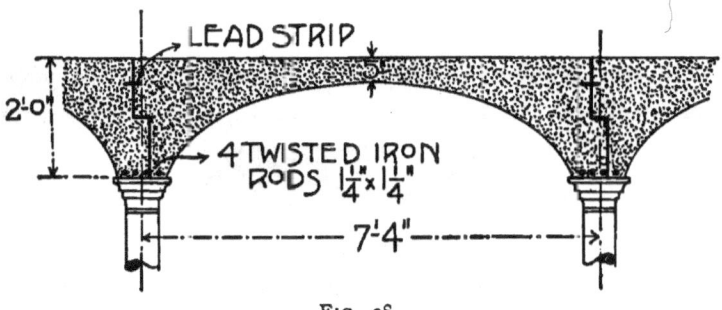

Fig. 38.

on twisted iron rods $1\frac{1}{4}'' \times 1\frac{1}{4}''$ for support (see Fig. 38). The concrete arches are 7' 4" centre to centre of columns without the aid of any metal-work. The columns are placed 25' 0" centres longitudinally, with 4 twisted rods acting as supports between. The concrete slabs are joined by a lap joint, with a lead strip embedded to prevent the passage of water. This type was tested to 300 lbs. per square foot. The arches deflected $\frac{1}{8}''$ at the centre and remained uninjured. Such construction is hardly applicable to office buildings on account of the curved soffit, and small transverse space between columns, but modifications of this form would seem to offer many advantages in roof construction.

SEGMENTAL ARCHES OF TILE.

For long spans in buildings where a flat ceiling is not necessary, as in warehouses, etc., a segmental arch is often used, following the curve of pressure, as shown in Fig. 39.

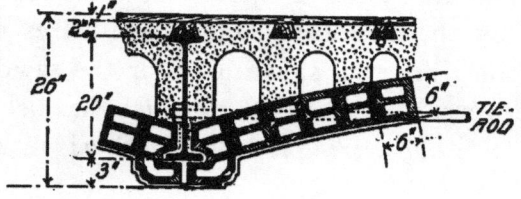

FIG. 39.

The tie-rods, spaced equally, are encased in tile to give a panelled effect. The arch shown in Fig. 40 was used with

FIG. 40.

extra heavy tiles in the Sibley Warehouse, Chicago.* The use of such segmental arches for office buildings has been abandoned after a trial in the Rand-McNally Building, Chicago. A ceiling of flat tile was there suspended under a segmental arch, but it did not prove successful, and has not since been used.

The floor of N. Poulson also deserves special notice, but the use of these particular arches seems somewhat limited up to the present time to public buildings, libraries, etc., where the groined arch is more suitable than in office structures. There is no example of the Poulson arch in Chicago, to the writer's knowledge. The system may be described as follows: The total floor-space is divided into panels of about 25' each way by the columns, with con-

---

* Tests at Washington, D. C., March 26, 1894, on a segmental arch 15' 4" span, $\frac{1}{16}$" rise, and using blocks 8" at the haunches and 6" at the centre, developed a safe capacity of 1000 lbs. per square foot of bearing surface.

necting lattice girders. These panels are spanned by a system of arched flats, generally 3" × 1", with a rise of 18". The thrust of the arches is taken up by an octagonal frame of angle irons in each panel. All arch intersections are bolted. These flats are built into the lower parts of concrete beams, which carry the floor on their upper edges, and curved plaster soffits on the under sides, forming the ceiling ribs. A rubber bag, held up by an umbrella scaffold, is pressed up into the triangular space formed by the intersecting ribs, and a plaster of Paris or cement soffit is formed with the curved bag for support. Heavy steel wires are then stretched over the system, which wires, in turn, support galvanized wire cloth. A 3" cement filling is then placed on top to hold the nailing-strips.

"GUASTAVINO" ARCH.

The peculiar strength of the egg-shell, or of any continuous layer of material, flat, curved, or dished, like the buckle-plate for example, undoubtedly suggested the form of the Guastavino arch. Arch or dome shells are built of small rectangular tiles of hard terra-cotta, three or four layers being used, of 1" thickness each, laid together in either square or herring-bone bond. Portland cement is used for the joints and between the concentric layers or shells. The great strength of these arches lies in the fact that they follow closely the curve of pressure, thus avoiding tension in the voussoirs, and in the fact that the successive layers break joint so perfectly that to open any joint several tiles must be sheared off. The great disadvantage in the use of this type in mercantile or office buildings lies in the curved soffit, and the necessary use of exposed tie-rods where several spans occur side by side. In solid masonry construction, as in libraries, public buildings, etc., where the walls or piers are capable of resisting the horizontal thrusts, and where a curved soffit is in keeping, this type possesses great advantages.

It will be noticed that little has been said as regards the comparative cost of the types of flooring here mentioned. This question will undoubtedly serve as a prime factor in making a choice between the various methods, but, as stated before, the question of expense should be held entirely subservient to that of safety, both present and future. Two different types of floor construction, with a considerable variance in the ultimate capacity, cannot properly be compared in the question of cost. If *all* methods meet the maximum requirements, the conditions are equal, and the cost may be considered as the determining factor. The following table gives the comparative costs of the hollow-tile and Melan floors.*

Material.	Hollow-tile Floor (see Fig. 30). Total load = 150 lbs. per sq. ft.	Melan. Total load = 150 lbs. per sq. ft.	
		6' 8" Span (see Fig. 33).	20' Span (see Fig. 34).
	Cts.	Cts.	Cts.
Beams, connections, tie-rods, etc....	11.44 lbs. @ 3 c. = 34.3	10.12 lbs. @ 3 c. = 30.4	4.95 lbs. @ 3 c. = 14.9
Bent shapes......	....	1.8 lb. @ 3½ c. = 6.3	2.25 lbs. @ 3½ c. = 7.9
Arching..........................	26	14	20
Levelling....	2	4	7
Depth............................	16" @ 2 c. = 32	16 " @ 2 c. = 32	13 " @ 2 c.= 26
Flat ceiling	....	6	
Cost, cents per sq. ft...., ...	94.3	92.7	75.8

CHICAGO BUILDING LAWS—FLOOR ARCHES.

The following requirements are specified in the Chicago building ordinance, Section 117: "The filling between the individual iron or steel beams supporting the floors of fire-proof buildings shall be made of brick arches, or concrete arches, or hollow-tile arches, or Spanish tile arches. Brick arches shall not be less than 4 inches thick, and shall have a rise of at least 1¼ inches to each foot of span between the beams. If the span of such arches is more than 5 feet, the thickness of the same shall not be less than 8 inches. If

---

* See *Transactions Am. Soc. Civil Engineers*, vol. xxxi. No. 4.

hollow-tile arches having a straight soffit are used, the thickness of such arches shall not be less than at the rate of 1¼ inches per each foot of span. If Spanish tile arches are used, they are to be made as per the published formulæ of the Guastavino Construction Company, subject to the verification and approval of the Commissioner of Buildings. If concrete arches are used, the concrete in the same shall not be strained more than 100 pounds per square inch, if the concrete is made of crushed stone, nor more than 50 pounds per square inch, if the concrete is made of cinders. In all cases, no matter what the material or form of the arches used, the protection of the bottom flanges of the beams and so much of the web of the same as is not covered by the arches shall be made as before specified for the covering of beams and girders."

Again, Section 88: "Hollow tile and porous terra-cotta may be used in the form of flat arches for the support of floors and roofs; such floor arches having a height of at least 1¼ inches for each foot of span. The arches must be so constructed that the joints of the same point to a common centre; the butts of the arches shall be carefully fitted to the beams supporting them; and there shall be a cross-rib for every 6 inches or fractional part thereof in height; and in addition to these there shall also be diagonal ribs in the butts. Floor arches made in the form of a segment of a circle or ellipsis must be constructed upon the same principles, but in such cases the individual voussoirs forming the arch shall not be less in height than one thirtieth of the span of the arch. Such arches, whether flat or curved, shall have their beds well filled with mortar, and the centres shall not be struck until the mortar has been set."

Before leaving the subject of fire-proof floors it will be well to mention the test of hollow-tile arches provided in the case of the Chicago Athletic Club Building, before mentioned. The steel beams where not fire-proofed were badly bent where the ends were not held, but the metal portions were in perfect condition where the fire-proofing remained intact. Not a single floor arch fell, and "tests since made on the worst-looking ones have developed a sustaining capacity of 450 lbs. per square foot without sign of rupture."

The arches treated of in this article, as affecting the method of design of the floor-beams, girders, and columns, are the ordinary tile arches—be they of the older Pioneer construction, the Lee form, or the newer arches similar to the Johnson type.

FLOOR LOADS.

Before considering the most economical arrangement of floor-beams, the question of loads, which will largely govern the design of the floor system, must be examined. The loads in building construction may be classified as dead, live, wind, and eccentric loads. These will all be considered in their proper places in these pages. The principal loads affecting the floor system are:

Dead Loads, comprising all of the static loads due to the constructive parts of the building, stationary machinery, water-tanks, and any other permanent loads.

Live Loads, comprising the people in the building, office furniture, movable stocks of goods, small safes (large safes require special provision), or varying loads of any character.

The maximum live load per square foot is usually assumed as follows:

For crowd of people	80 lbs.
For floors of houses	40 "
For theatres and churches	80 "
For ball-rooms or drill-halls	90 "
For warehouses, etc.	from 250 " up.
For factories	200 to 450 lbs.

While 80 lbs. is the maximum possible live load per square foot from a crowd of people (unless dancing be considered), still we can hardly expect to realize any such load under the conditions governing an office building. Large crowds very seldom collect in offices, except, perhaps, on the two or three lower floors devoted to stores or banking purposes, and greater allowances are generally made for such places. The ordinary office furniture will certainly not exceed, and seldom equal, the weight allowed for persons, and hence additional security is introduced. Prof. Baker, in his "Treatise on Masonry Construction," gives 10 lbs. per sq. ft. movable load for dwellings, 20 lbs. for large office buildings, 100 lbs. for churches, theatres, etc., and from 100 to 400 lbs. for stores, warehouses, and factories, according to contents.

A 20-lb. unit load in office buildings, as recommended by Prof. Baker, might be seriously questioned, and late experiments in this direction would seem to sustain the criticism. While 20 lbs. per square foot *may* be amply sufficient for average loads at present, we must remember that the use of an average is always dangerous, while provision *should* be made, but not recklessly, for all possibilities of extremes, either present or future. An article in the *American Architect*, August 26, 1893, gives the results of some experiments made by Messrs. Blackall & Everett, Boston architects, on the actual weights of all moving loads in some of the larger Boston office buildings. The loads considered were those due to people and all possible movable articles, including all office fittings except such as were a part of the floors or partitions, radiators excepted. The results were as follows: In 210 offices in the Rogers, Ames, and Adams buildings, an average of 16.3 lbs. per sq. ft was found for the Rogers Building, 17 lbs. for the Ames and 16.2 lbs. for the Adams Building. The *greatest* moving load in any one office in the three buildings was 40.2 lbs. per sq. ft., while the *average* for the heaviest ten offices in each of these buildings was 33.3 lbs. per sq. ft. Mr. Blackall concludes: " If these figures are to be trusted to any extent whatever, then even under the most extreme circumstances, taking the pick of the heaviest offices in the city and combining them into one tier of ten stories, the average load per square foot would be only a trifle over 33 lbs., while for all purposes for strength, an assumption of 20 lbs. would be amply sufficient in determining the loads on the foundations, as well as on the columns of the lower stories."

With a proper provision, then, for maximum loads in the floor system, the 20 lbs. recommended by Prof. Baker is not enough, though safe, perhaps, for an average. But, as remarked before, the use of averages is dangerous, and it becomes a very nice problem to balance present economy with maximum present requirements or future possibilities; for the present weight per square foot may not safely be taken as the maximum occurring during the life of the building. The municipal laws of New York and Boston provide for a moving load of 100 lbs. per sq. ft., while those of Chicago require 70 lbs. live load per. sq. ft. on the floor system, with proper reductions for the columns and footings. With a proper regard for economy 100 lbs. per sq. ft. would certainly seem too large; 80 lbs. for the lower and busier floors, and 40 lbs. for the upper or office floors, are certainly safe, and good averages, considered in all lights. These loads, used in all calculations affecting the metal framing, must not be confounded with the required loads for the strength of the individual tile arches. While the live load per square foot may be reduced over large areas in proportioning the metal-work, the maximum possible live load must still be used when any single floor arch is considered by itself, or subjected to tests to determine its strength. Some further data on live loads are given later under a discussion of the building laws of New York, Boston, and Chicago.

In the Mills Building, erected in San Francisco in 1891, the live loads were as follows:

	Beams.	Girders.	Columns.	Footings.
First floor	60	50	40	—
Second floor to attic.	40	30	20	—
Roof	20	15	10	—
Rotunda	60	50	40	—

The practice in Chicago seems to be pretty well defined in the matter of decrease of live loads per sq. ft., as they are transferred from beams to girders, from girders to columns, and thence down the columns to the footings. This practice is founded on the supposition that it is quite possible that the beams may sometime have to carry their full capacity in live loads, while the chances are increasingly less that the girders or columns will ever be required to carry anywhere near their full capacity, if a full load had been assumed. The fully loaded area would probably never be large, and a girder or column would rarely, if ever, lie in the centre of such an area. The effect of a live or moving load, causing vibration in the parts of the structure, is also gradually lessened, as the vibration is taken up in the transfer of the load from member to member, so that by the time it reaches the footings or foundations the live load is ignored entirely. In fact, we can hardly imagine the perceptible effect on the foundations of the people in an office building, as compared with the infinitely greater *dead* load, due to the structure itself.

In the Venetian Building in Chicago the beams were calculated for the following live loads:

Upper floors...... .... 35 lbs.
Second, third, and fourth floors........ 60 "
First floor........................... 80 "

Girders carry 80 per cent, columns 50 per cent.

The dead loads to be considered in the floor system include the arch itself, beams, concrete filling, floors (wood, marble, or mosaic), ceilings, and partitions.

The weights of the iron or steel beams and tile partitions are actually calculated for a typical floor plan, and then rated at so much per sq. ft. of floor surface. This is absolutely necessary in regard to partitions in office buildings, as they are constantly being changed to suit the convenience of tenants. The weight of the arch varies with the depth; the depth is dependent on the span. In the annex of the Marshall Field Building, Chicago, the following weights were used:

Flooring, ¾-inch maple............ ... 4 lbs.
Deadening....................... 9 "
15-inch tile arch................... 45 "
Iron............................. 12 "
Plaster.......................... 5 "
Partitions, 3-inch mackolite......... 20 "
Total...................... 95 lbs. dead load.

We have, therefore, for live and dead loads as follows:

	Beams.	Girders.	Columns.	Footings.
Offices: Live..........	85	65	45	....
Dead...........	95	95	95	95
Total.......	180	160	140	95
Store floors: Live......	95	75	55	....
Dead......	95	95	95	95
Total...	190	170	150	95

The dead loads assumed in the Old Colony Building, Chicago (1893), comprised:

Flooring	4 lbs.
Deadening	18 "
Tile arches	35 "
Iron	10 "
Plaster	5 "
Partitions	18 "
Total	90 lbs.

The dead and live loads used in the calculations of the floor systems and columns of this building were, in pounds per square foot:

	Beams.	Girders.	Columns.	Footings.
Live	70	50	40	—
Dead	90	90	90	90
Total	160	140	130	90

The floors for the Fort Dearborn Building were calculated in accordance with the following data:

	Dead Load		Live Load.	
	Beams.	Girders.	Beams.	Girders.
1st floor	85	85	125	110
2d to 13th floors	75	75	70	60
Roof	40	40	40	40
Sidewalk	120	140	200	180
Prismatic lights	50	50	200	180
Skylight	....	....	40	40
Stairs	50	50	70	60

The live load on the beams from the second to thirteenth floor inclusive was taken at 70 lbs. per sq. ft., and an additional load of 20 lbs. per sq. ft. was added to the dead load to care for all partitions which were likely to be moved at any time.

The girders were figured for partition loads at 20 lbs.

per sq. ft. for all movable partitions, and for the actual loads of the main partitions.

The live load on the columns was taken at 50 lbs. per sq. ft. from the second to the twelfth floor inclusive, plus the girder reactions for partitions.

The following table gives the unit loads used in figuring the columns:

	Live Load on Floor.	Live Load on Columns from Floors above.	Total Load on Columns.
Roof.......	40	....	....
13th floor...	50	40	40
12th " ...	....	45	85
11th " ...	....	41	126
10th " ...	....	35	161
9th " ...	....	31	192
8th " ...	....	25	217
7th " ...	....	21	238
6th " ...	....	15	253
5th " ...	....	11	264
4th " ...	....	5	269
3d " ...	....	1	270
2d " ...	....	0	270
1st " ...	125	0	270
Basement...	....	50	320

The dead load on the floor-beams was made up as follows, a 9" porous end-construction arch having been used:

9" arch........................	26	lbs. per sq. ft.
9" 21-lb. I beams...............	4	" " " "
6 to 1 cinder concrete...........	30	" " " "
Mosaic and wood floors, average..	10	" " " "
Plaster........................	6	" " " "
Total...................	76	" " " "

FLOOR FRAMING—BEAMS.

The distance centre to centre of the floor-beams must be determined with reference to the type of floor arch used. Ordinary practice in Chicago skeleton construction has made from 5 to 6 feet the usual span for tile arches, in panels of ordinary lengths; but in cases where the columns are spaced a considerable distance apart the floor-beams are placed nearer together. Reference to Figs. 18 and 19 will show the practice in beam-spacing in late examples of skeleton buildings in Chicago.

The most economical arrangement of floor-beams has had little investigation, and there seems to be no uniformity of practice. If the framing plans could be so arranged that the floor-beams and girders would be strained to the full allowable fibre strain, it would certainly be more economical than where the framing plans require the use of beams heavier than those actually needed. Take, for example, a framing plan calling for a bending moment in a floor-beam of 65,000 foot-pounds. This would require a moment of resistance of 48.72. The moment of resistance for a 12" 40-lb. beam is only 46.9, while R for a 15" 41-lb. beam is 56.6. The latter would have to be used, with an excess in strength of some 16 per cent; and if such panels occurred frequently in a floor system, an excess of 16 per cent would therefore occur throughout. Hence an economical framing plan would be one in which the beams are so arranged in span and distance centre to centre, as to carry a given floor load with the beams strained to the full allowable fibre strain. A very small variation may make this possible or impossible.

Again, it is seldom economical to use the heaviest weight of any depth of beam, if a deeper beam can be used. There is necessarily a great waste of material toward the ends of heavy rolled beams, and as the strength increases as the square of the depth, the deeper beam is always the more economical. Thus the moment of resistance for a 12" 32-lb. beam is 37, while the 10" 33-lb. beam has $R = 32.3$. The former is lighter, and far stronger. A 20" 64-lb. beam is also stronger than a 15" 80-lb. beam.

The coefficient for a 15″ 50-lb. beam is 753,000.
"   "   " " 15″ 55-lb. "   " 792,000.
"   "   " " 15″ 60-lb. "   " 916,300.

Hence the use of a 15″ 55-lb. beam is not economical, as the coefficient does not vary between the 50- and 60-lb. limits in proportion to the weight. For a uniformly distributed load these coefficients are obtained by multiplying the load, in pounds uniformly distributed, by the span length in feet. If the load be concentrated at the centre of the span, multiply the load by 2, and then consider it as uniformly distributed. The maximum coefficients of strength for I beams of different depths and weights are usually given in the pocket companions issued by the various steel companies. The handbook of Carnegie, Phipps & Co. is generally used by architects and engineers. In that book the maximum permissible coefficients are given for all of the ordinary rolled shapes, on a basis of 16,000 lbs. per square inch fibre strain, and also on a basis of 12,500 lbs. per square inch fibre strain. The former is generally used in building work.

The distribution of the material in the cross-section affects the moment of inertia, and hence R. The sections from some mills will be found better than those from others in this respect.

Care must be taken in figuring floor-beams to see that the length of clear span is not too great, giving a deflection sufficient to crack the plaster ceiling beneath. A deflection of about $\frac{1}{360}$ of the clear span, or $\frac{1}{30}$ of an inch per foot, has been found by experiment and practice to be the maximum permissible deflection—or $\delta = L \times 0.33$, where δ = greatest allowable deflection in inches, at centre of beam, and L = length of span in feet. This safe deflection limit is also indicated for each size and weight of beam given in the tables for uniformly loaded I beams in the handbook of Carnegie, Phipps & Co.

Lateral stiffness may also need consideration in some cases. Where the floor-beams are of the same depth as the girders, "coping" is necessary, or a cutting away of the ends of the floor-beams to fit against the flanges of the girders. About ¼ inch clearance is usually allowed between floor-beams and girders, and ¼ inch between columns and girders. This is sufficient for easy erection.

The standard connection-angles manufactured by Carnegie, Phipps & Co. are generally used whenever practicable, as connections between floor-beams and girders. These connection-angles are given for the various depths and weights of steel and iron beams in the handbook. They are designed on a basis of 10,000 lbs. allowable shearing-strain, and 20,000 lbs. bearing on rivets or bolts per square inch, and are *usually* of sufficient strength for regular details as found in practice. The adoption of such

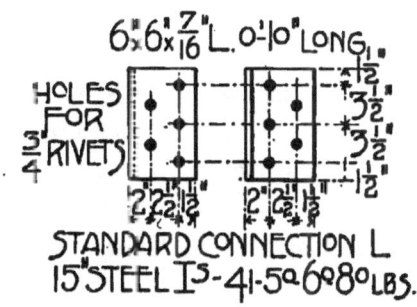

FIG. 41.

uniform "standards" is certainly a great help to the mills and bridge or iron shops, as well as to the designer, but in the hands of the careless or ignorant designer is apt to be an element of weakness. From careful observation of building methods as practiced in Chicago, the writer is convinced that faulty details constitute an even greater part of the defects in the general run of buildings, than arises from poor materials employed, or imperfect general features of design. Any "standards" are therefore to be used with caution, as they tempt the careless designer to use them under all conditions, whether they be adequate or not. They are *standard*, hence they *must* be all-sufficient.

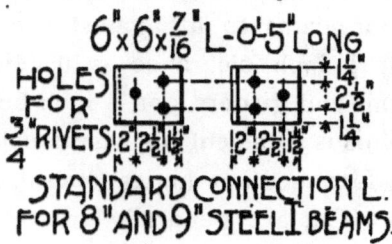

FIG. 42.

Figs. 41 and 42 show standard connection-angles for the beams as given.

GIRDERS.

The girders, running from column to column, support the floor-beams, and transfer their loads directly to the columns. As before mentioned, it is often necessary to use two I beams side by side as a girder, or even plate or latticed girders in longer spans or under special loads. Separators should always be used in the case of double beams, in order to equalize the loads on the two beams, and also to act as spacers, keeping them a proper distance apart. Carnegie's separators are generally taken as standard.

It is quite impracticable to make any comparisons as to the relative economy of short spans for girders with many columns, and fewer columns with longer girders. Both types are to be found in Chicago, even to extremes, but they are usually the results of *conditions*, rather than attempts at economy. The conditions governing the design of any particular building are usually so potent that the rule in one case might prove the exception in the next. The arrangement of the exterior piers, the architectural effect striven for, the arrangement and proper planning of the interior for the uses intended, all govern, in a great measure, the placing of the supporting columns, and hence the girder lengths. Thus in Fig. 18, showing the framing plan of the new Fort Dearborn Building, two 9-inch beams or two 10-inch beams are generally used as girders, while in Fig. 19, of the Reliance Building, single beams are used as girders in all cases. In the Woman's Temple, Chicago, Burnham & Root, architects, the floor-beams are nearly all 30 feet long, and the girders likewise, 15-in. I beams having been used throughout; while in the new Marshall Field Building, by the same architects, a 20-foot panel was used.

In office buildings the panels are often made of such dimensions as to give two suitable office widths from centre to centre of piers. Thus the practice of Holabird & Roche, architects, is to space the exterior and interior columns 23 feet centres where possible, making two offices of 11 ft. 6 in. in each bay (see Fig. 16).

A point to be remembered in the design of girders is that a much more economical girder can be had when two floor-beams are to be supported than three; or an even number instead of an odd number of beams. In the latter instance a load will occur at or near the centre of the girder, resulting in a much greater bending moment. If but two beams are used, the arm is but one third the span of the girder. All of the floor-beams and girders in the floor system are usually so arranged as to be flush on the under sides, as shown in Fig. 43. This is to provide for the plastered ceiling. The inequalities in the arch depths are made up in the concrete filling.

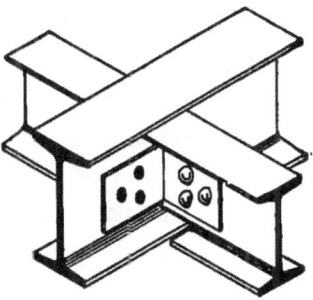

FIG. 43.

CHAPTER V.

EXTERIOR WALLS—PIERS.

THE subject of the exterior piers which carry their tributary floor and roof loads, besides the weight of the walls themselves, is capable of three separate treatments, each of which is used under its own peculiar circumstances.

First. Where the outside piers are constructed entirely of masonry, carrying all of the wall-, floor-, and roof-loads which come on them, by means of masonry alone. Such construction is used in buildings of moderate height, and constitutes the ordinary type of building. But in the higher structures of from sixteen to twenty stories, which are here being considered in particular, it is the rare exception, at the present time, to rely entirely on masonry piers.

The objections to such piers of solid masonry are three-fold:

a. The modern requirements of plenty of light and air in all offices, demand that the windows be broad and numerous and the piers narrow. In the highest buildings of the present day hardly any masonry construction is strong enough to carry the necessary roof- and floor-loads besides its own weight, for so great a height and with so small a cross-section as is desired. There are prominent office buildings in Chicago and elsewhere, in which the exterior walls carry their proper share of all loads; but a little observation will show that in high buildings of this type the comforts of the tenants have, in a large measure, been sacrificed for architectural effect.

b. The second objection to such large masonry piers is

that they take up too much valuable renting-space. When the rent of offices is proportioned at so much per square foot, this becomes a matter of no inconsiderable importance to the owner.

c. The weight of these solid masonry piers would so add to the load per square foot on the clay or foundations that many of the most remarkable examples of architectural engineering would be well-nigh impossible.

In the new Marshall Field Building in Chicago, masonry piers were used to carry all exterior loads, but a mercantile structure does not present as exacting conditions as an office building, and the exterior piers may be widened for architectural effect without seriously inconveniencing the plan of the interior.

Second. The second treatment of which the exterior piers are capable is that in which metal columns, carrying the tributary floor and roof loads, are placed inside the masonry piers, while the latter support themselves and the "spandrels" only. The spandrels constitute those portions of the exterior walls lying between the piers and over and under the window-spaces.

If this method is employed, great care must be taken that the masonry does not touch the columns, in order that the *unequal* settlement of the metal-work and the masonry may not cause undesirable strains. On account of the numerous mortar-joints, the masonry will settle much faster than will the metal columns under the gradual settlement of the whole structure. As an example of initial compression in freshly laid mortar, Mr. Geo. B. Post, architect of the New York Produce Exchange building, states that a measured height of 9' 6" at the time of building, compressed about $\frac{1}{2}$" under a maximum pressure of 62 lbs. per square inch of base, induced by the finished wall. The whole wall was built very rapidly.

If, then, the masonry bears on rivet-heads, plates, or connections on the columns, a heavy strain is produced which has not been provided for. Great care is necessary in such combinations of metal columns and masonry piers to leave sufficient "open joints" at points over cornices and the like, where they will least be noticed, to allow for such settlement. Also where the mass is not homogeneous, as in stone facing and brick backing, the result is likely to be that the stone, with fewer mortar-joints, settles less and receives more than its share of the load, thus producing cracks and spalling off the angles. This was the case in the old portion of the Washington Monument.

The objections of size and weight will also hold in the piers of this type, as in the first method, if the building be very high. Thus in the Masonic Temple of twenty-one stories, metal columns of plates and angles were placed within the masonry piers, but it was found that the maximum allowable pressure of 12 tons per square foot on brickwork, as used by the engineer, would be reached at the level of the fifth floor; hence below that level the load exceeded the safe compressive resistance of the material; and this without any floor or roof loads, as the latter were carried by the metal columns within the piers. The expedient was therefore adopted of carrying the masonry-work on brackets attached to the metal columns at the sixteenth- and fifth-story levels, thus making the pier consist of three separate columns of masonry, and the one continuous metal column.

Third. The third method of constructing the exterior piers is the one more approved at the present stage of architectural engineering—the one which has undoubtedly opened up the means for building the highest structures. In this, *all* weights are thrown on the metal columns, which, in place of solid piers, are surrounded with a protective shell or covering only, made of ornamental terra-cotta or brickwork, securely anchored to and supported by the columns at the various floor levels.

This construction undoubtedly gives the minimum weight per foot of height, and makes possible such small piers as are indispensable for light and desirable offices. The "Chicago type" is a popular name for this method; a type which has developed fast and remarkably during the past ten years of Western architecture, while the height of municipal buildings has been increasing steadily from ten to twenty stories. The increasing value of ground-space, the demands for rapid construction, and the necessity for the lightest possible loads on the subsoil, have all contributed to the success of this type.

Chicago construction thus does away with masonry as a supporting member, and the load-bearing brick wall or masonry pier is replaced by an envelope of terra-cotta or brickwork, enclosing the steel columns and filling the spandrels or spaces between the windows. This envelope is not used as a strengthener to the supporting members, but as a protection against the elements and the dangers of fire. The brick wall, once the fundamental factor in building construction, now fulfils simply a decorative and protective function. The great possibility for external effect through this use of brick and terra-cotta in connection with skeleton construction, has opened up a vast market to the manufacturers of fine qualities of face-brick, moulded brick, and terra-cotta in all its varieties.

The terra-cotta companies design their pieces with special reference to tying them to, or suspending them from such a framework; so that, in reality, the building becomes nothing more nor less than a vital skeleton of steel, with an architectural and protective wrapper of terra-cotta, tile, or brickwork, inside and outside. The terra-cotta arches, which to the casual observer seem to carry some heavy wall or pier above, prove to be made of hollow clay blocks, held by wires or clamps to the concealed beams or girders which really support the loads.

Brick and terra-cotta are generally preferred to other building materials for the exterior walls of a tall building, on account of the ease with which they may be handled as well as for the facility with which they may be built into and about the forms of beams and columns. Stone has gradually been driven from the field of skeleton construction in exterior walls, except as used in the lower stories only, as a base for the superimposed brick or terra-cotta work. This has been due to the difficulty experienced in properly attaching the masses of stone to the metal framework. Stone has also been used in thin slabs in the lower stories, as in the first floor of the Reliance Building, where highly polished slabs of granite were enclosed in ornamental frames or grilles of metal-work surrounding the columns, as shown in lower part of Fig. 44. Fig. 45 shows the girder over the main entrance to the Masonic Temple.

In order to render the exterior impervious to moisture, and thus protect the metal framing against corrosion, only the very hardest and most thoroughly burned brick should be used. Portland cement mortar is also specified in the best classes of work, with well-filled joints and careful bonding and anchoring. In other words, less is now required of the brick wall as a *supporting* member than formerly, when the walls fulfilled the function of bearing dead loads only; but much more is now demanded of it as to quality and perfection of workmanship, and hence a better constructed and more thoroughly knit wall has resulted in the best examples of Chicago construction.

The Chicago building ordinance defines skeleton con-

Fig. 44.—Detail of Terra Cotta, Reliance Building.

struction as follows: "The term 'skeleton construction' shall apply to all buildings wherein all external and internal loads and strains are transmitted from the top of the building to the foundations by a skeleton or framework of metal. In such framework the beams and girders shall be riveted

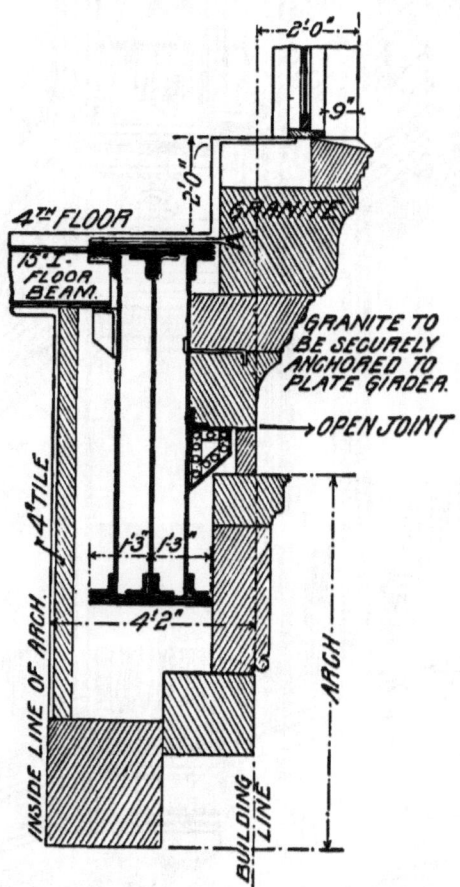

Fig. 45.—Section over Main Entrance, Masonic Temple.

to each other at their respective junction points. If pillars made of rolled iron or steel are used, their different parts shall be riveted to each other, and the beams and girders resting upon them shall have riveted connections to unite them with the pillars. . . . If buildings are made fire-proof entirely, and have skeleton construction so designed that their enclosing walls do not carry the weight of floors or roof, then their walls may be reduced in thickness one third from the thickness hereinafter provided for walls of buildings of the different classes, excepting only that no wall shall be less than 12 inches in thickness; and provided, also, that such walls shall be thoroughly anchored to the iron skeleton; and provided, also, that wherever the weight of such walls rests upon beams or pillars, such beams or pillars must be made strong enough in each story to carry the weight of wall resting upon them without reliance upon the walls below them. But if walls of hollow tiles are used as filling between the members of the skeleton construction, they shall be of the full thickness specified for non-skeleton buildings."

The requirements for protecting external structural members of iron and steel are defined as follows: " All iron or steel used as a supporting member of the external construction of any building exceeding 90 feet in height shall be protected as against the effects of external changes of temperature and of fire by a covering of brick, terra-cotta, or fire-clay tile, completely enveloping said structural members of iron and steel. If of brick, it shall be not less than 8 inches thick. If of hollow tile, it shall be not less than 6 inches thick, and there shall be at least two sets of air-spaces between the iron and steel members and the outside of the hollow-tile covering. In all cases the brick or hollow tile shall be bedded in mortar close up to the iron or steel members, and all joints shall be made full and solid.

"Where skeleton construction is used for the whole or part of a building, these enveloping materials shall be independently supported on the skeleton frame for each individual story.

"If terra-cotta is used as part of such fire-proof enclosure, it shall be backed up with brick or hollow tile; whichever is used being, however, of such dimensions and laid up in such manner that the backing will be built into the cavities of the terra-cotta in such manner as to secure perfect bond between the terra-cotta facing and its backing.

"If hollow tile alone is used for such enclosure, the thickness of the same shall be made in at least two courses, breaking joints with and bonded into each other."

The New York law prescribes the following: "Where columns are used to support iron or steel girders carrying curtain-walls, the said columns shall be of cast iron, wrought iron, or rolled steel, and on their exposed outer and inner surfaces be constructed to resist fire by having a casing of brickwork not less than 4 inches in thickness and bonded into the brickwork of the curtain-walls, or the inside surfaces of the said columns may be covered with an outer shell of iron having an air-space between; and the exposed sides of the iron or steel girders shall also be similarly covered in and tied and bonded."

The first example of a purely skeleton construction in Chicago occurred in the rear wall of the Phenix Building, now the Western Union Telegraph Building, by Burnham & Root, architects. In the wall behind the elevators, cast columns were used with two sets of horizontal supports at each story. The outside supports were made of I beams resting on brackets connected to the columns, these I beams carrying a $4\frac{1}{2}$-inch wall of enamelled brick. The inner supports consisted of I beams placed between the columns, supporting a 4-inch wall of hollow tile. Thus the wall was formed of two layers or "skins" held together by the window-frames, etc. To Mr. W. L. B. Jenney belongs the credit of having designed the first skeleton building erected in Chicago; the Home Insurance Building, built in 1883. This structure also contained the first Bessemer steel beams used in building construction.

To avoid any injury to the walls or piers in skeleton construction through the expansion and contraction of the tall columns of steel, the masonry or envelope must be so constructed as to be independent for each story length. This is provided by means of shelf-angles or brackets at each and every floor level, thus allowing the entire front of the building to be built in such a manner that any or all of the envelope or masonry facing may be removed without injury to the load-bearing members. In the Home Insurance Building just mentioned, cast lintels were used to form the soffits of the windows at each floor, and designed to carry the walls for the story above.

Fig. 46 shows a corner pier from the Reliance Building,

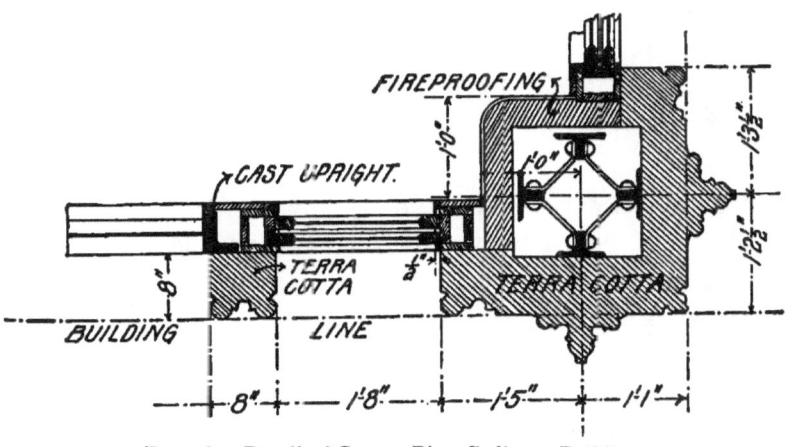

Fig. 46.—Detail of Corner Pier, Reliance Building.

and Fig. 47 is a plan of the supporting framework for same.

A striking example of what has been made possible in the construction of exterior piers by skeleton methods is shown in the difference between the old and new portions of the mammoth Monadnock Building in Chicago (see frontispiece). At the time of designing the older portion of this building, the owner, in spite of the protests of the architects, insisted on having the conservative practice of solid masonry piers, which, for a height of sixteen stories, resulted in walls some 6 feet thick at the street level. A few years later an addition was designed for the south half of the block, seventeen stories in height, and the walls of this new building were built in the veneer pattern, which

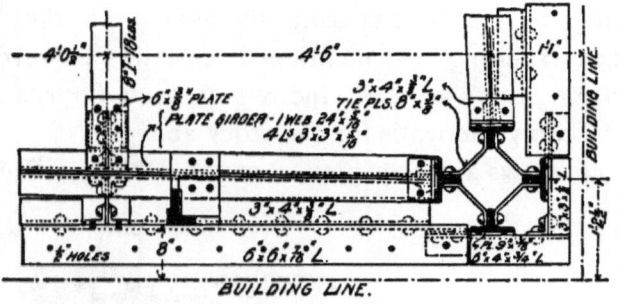

FIG. 47.

had previously been rejected by the owner of the other portion. It doubtless proved an expensive lesson for the first investor.

A brick wall carried to the height of the Manhattan Life Insurance Building in New York City (241') would, according to the building laws of most cities, have to be about 6 feet thick. Through the use of skeleton construction the enclosing walls in this building were made only 12 and 16 inches thick.

Fig. 48 shows the required thickness of walls under the Chicago ordinance for buildings devoted to the sale, storage, and manufacture of merchandise. Fig. 49, is for the walls of hotels, apartments, and office buildings of construction other than the skeleton type. Fig. 50 shows

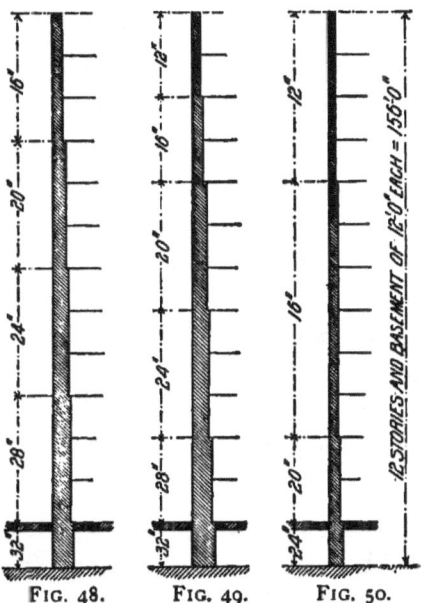

FIG. 48. FIG. 49. FIG. 50.

the requirements for masonry walls (in office buildings) which carry their own weight only.

CHAPTER VI.

SPANDRELS AND SPANDREL SECTIONS—BAY WINDOWS.

THE spandrels constitute those portions of the exterior walls, either on the street fronts or in the interior court, which lie between the piers and between the window-spaces of successive stories. "Spandrel sections," as they are called, must be made for every different type of spandrel support in the building, and they must clearly show the supporting beams or metal-work required to carry the veneer walls in the manner desired. These sections vary greatly, depending largely on the architectural effect contemplated by the designer in his arrangement of the material, and general descriptions of spandrels can hardly be given as applicable to general practice. Illustrations of numerous examples will better serve to show the methods employed.

The spandrel-beams are supported by the masonry piers where such load-bearing piers are used, or, in the veneer construction, by the metal columns in the walls. The face of the spandrel-walls may be "flush" with the piers, or "in reveal," that is, set back from the face of the piers. In the first case the wall presents a nearly unbroken surface, except for the terra-cotta sills and window-caps, while the second method accentuates the piers, and throws the spandrel-walls in reveal. The architectural treatment will determine these conditions. The former case is generally of far simpler construction, as the spandrel-beams come at or near the centres of the columns, thus avoiding many embarrassments in the irregular bracketing from the columns, which becomes necessary in the support of the spandrel-beams where the spandrel- or curtain-walls are recessed.

Fig. 51 shows a very simple form of spandrel section

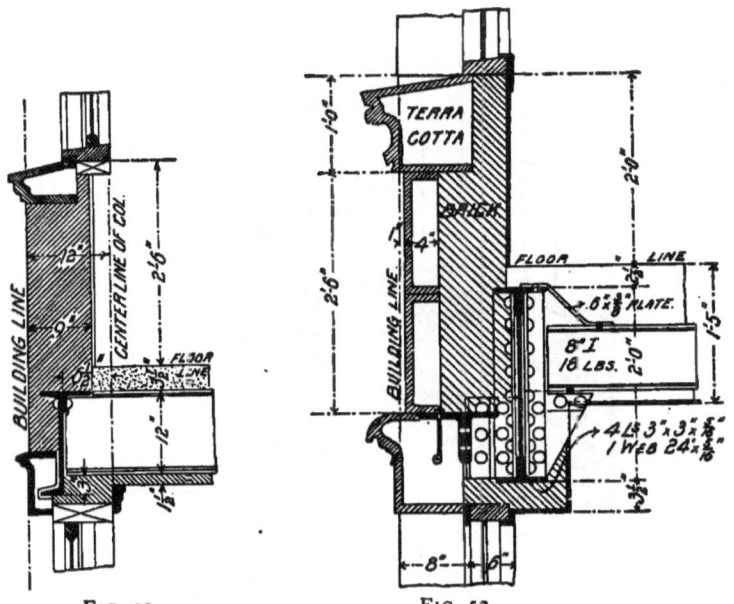

FIG. 51. FIG. 52.

from the Ashland Block, Chicago, where flush walls were used. The veneer wall is but 9 inches thick.

The use of plate girders, as the main spandrel supports, is shown in Fig. 52, which is a section taken from near the corner of the Reliance Building. The connections of these plate girders to the Gray columns used, are shown in Fig. 104, Chapter VII. The connections of the cast uprights to support the terra-cotta mullions between the windows, are shown in Fig. 53. Figs. 54 and 55 are taken from the eleventh- and twelfth-floor levels respectively of the Fort Dearborn Building. The section given in Fig. 56 is taken at the first-floor

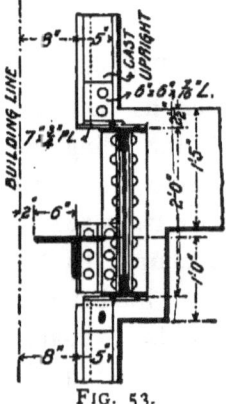

FIG. 53.

or sidewalk level, and shows the prismatic lights in the sidewalk, as well as the small windows which help to light the basement restaurant space. Fig. 57 is a section taken at the attic floor, showing the main cornice and roof construction.

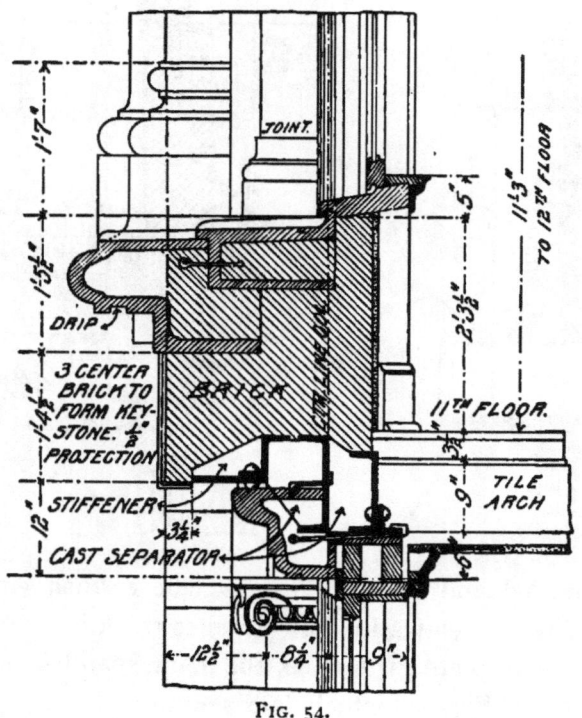

FIG. 54.

The materials generally used for veneer buildings consist, as before stated, of pressed brick and terra-cotta, the latter being used for the window-caps and sills, horizontal bands, ornamental capitals, brackets, etc., or even in entire fronts, as seen in the Stock Exchange Building, or in the Reliance Building of enamelled terra-cotta.

The brick or tile work of the piers is usually supported by bracket-angles, attached to the columns, as has been described in Chapter V, while the body or backing of the

SPANDRELS AND SPANDREL SECTIONS.

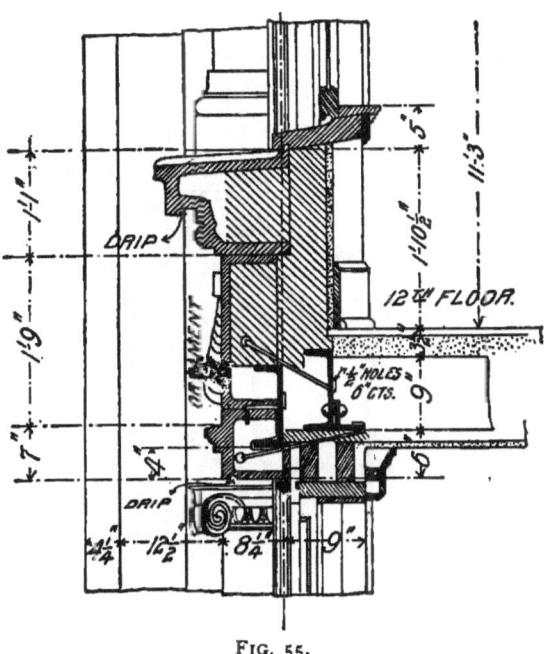

FIG. 55.

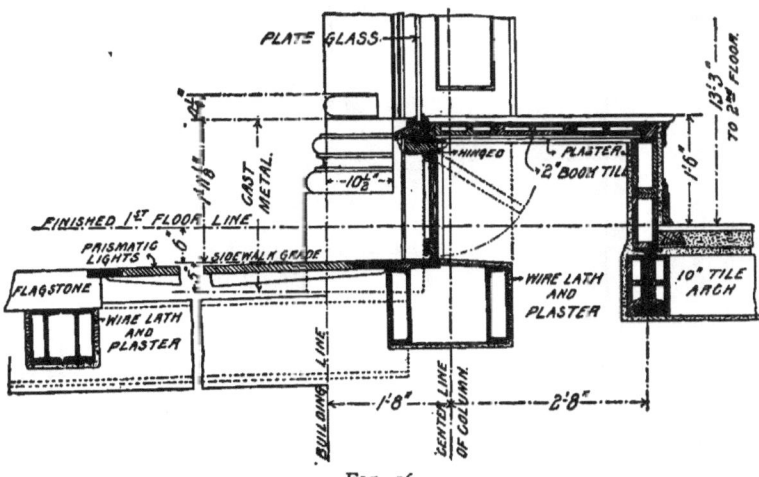

FIG. 56.

spandrel-walls is supported directly by the main spandrel-beams, as indicated in the previous figures.

The ornamental terra-cotta work, however, can seldom be supported directly by the spandrel-beams, and a system of anchors must be resorted to, to properly tie the individual

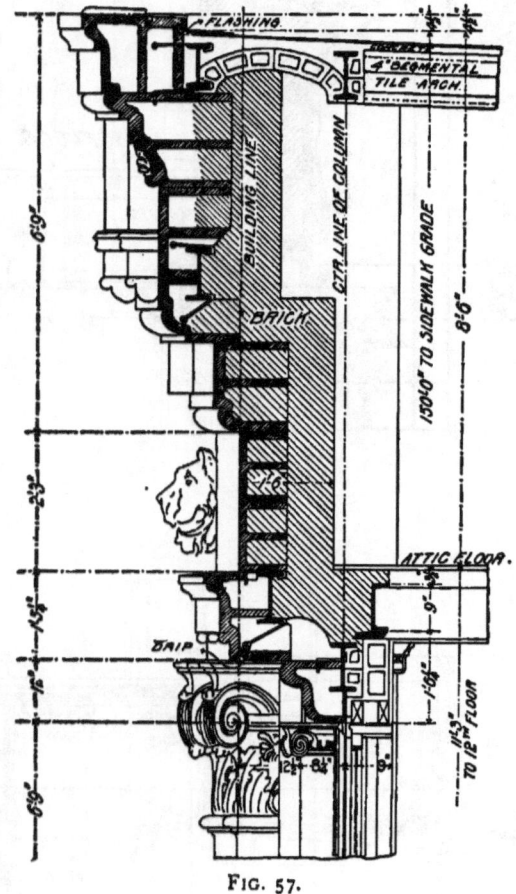

FIG. 57.

blocks either to the brick backing or to the metal-work itself. These anchors are usually made of ¼ inch square or round iron rods, which are hooked into the ribs provided in the terra-cotta blocks, and then drawn tight to the brickwork or metal-work by means of nuts and screw-ends.

Such anchors are shown in Fig. 59. Hook-bolts are also largely used, as in Fig. 55, where the ends are shown bent around the spandrel-channels or I beams. Clamps are frequently employed where the terra-cotta block lies snugly against a metal flange, as indicated in Fig. 59. The many possible methods which may be employed in securing proper anchorage cannot always be shown by drawings, and a proper execution of the work can only be

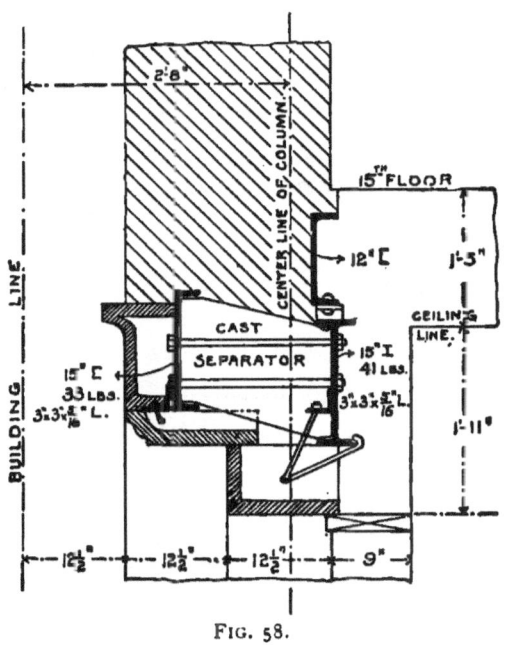

FIG. 58.

secured by most careful superintendence, and study in the field. The general scheme, however, must always be indicated on the sprandrel sections, as the holes necessary in the structural metal-work to receive the anchors should be included in the detail drawings of the iron or steel work, in order that such punching may be done at the shop.

Fig. 58 shows a sprandrel section from the Marquette Building, at the fifteenth-floor level. Heavy separators

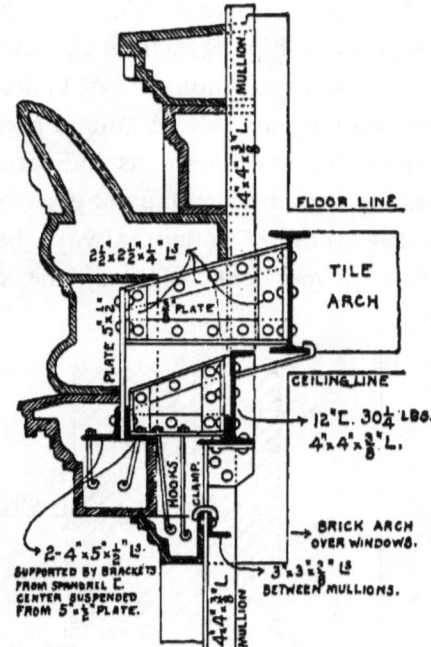

Fig. 59.

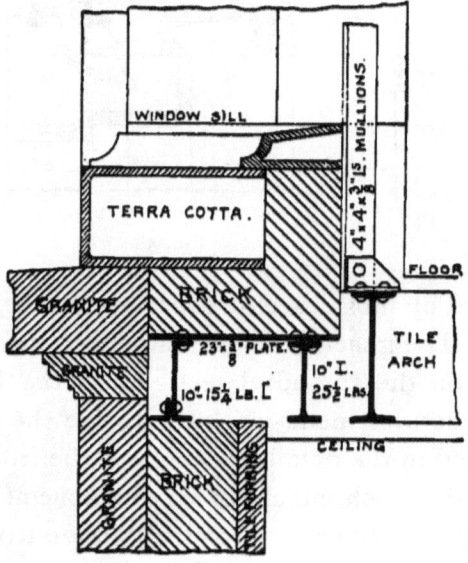

Fig. 60.

were used between the I-beam girder and the outside spandrel-channel.

A rather complicated spandrel section is that indicated in Fig. 59, taken from the Marshall Field retail store building. The spandrel-beams were here carried by the masonry piers used in the exterior walls. The section shown is taken where small ornamental balconies occur in the recessed wall between the piers. The vertical mullion-angles are plainly shown.

Fig. 60 is from the same building, taken at the level where the granite facing stops and the brick and terra-cotta work begins.

COURT WALLS.

The spandrel sections of the court walls differ in no way, as far as general principles are concerned, from those of the exterior walls. They are generally simpler, however, due to the plainer character of the wall, and to their usual decrease in thickness as compared to the exterior walls. A glazed brick is commonly employed, to reflect all possible light, while the sill-courses, etc., are of terra-cotta as before.

A section of the court wall in the Marshall Field Building is given in Fig. 61.

A simple court-wall spandrel section is shown in Fig. 62.

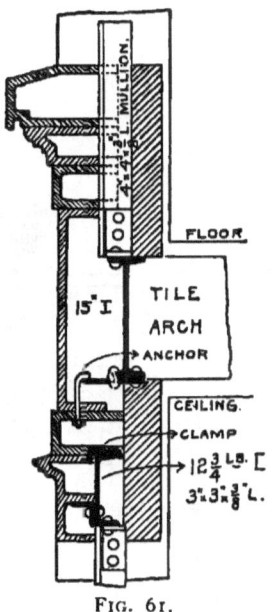

FIG. 61.

BAY WINDOWS.

With the introduction of the steel construction came the possibility and demand for the bay window, a feature which has certainly become very prominent in modern office-building and hotel design.

As in the ordinary spandrel section, the material for

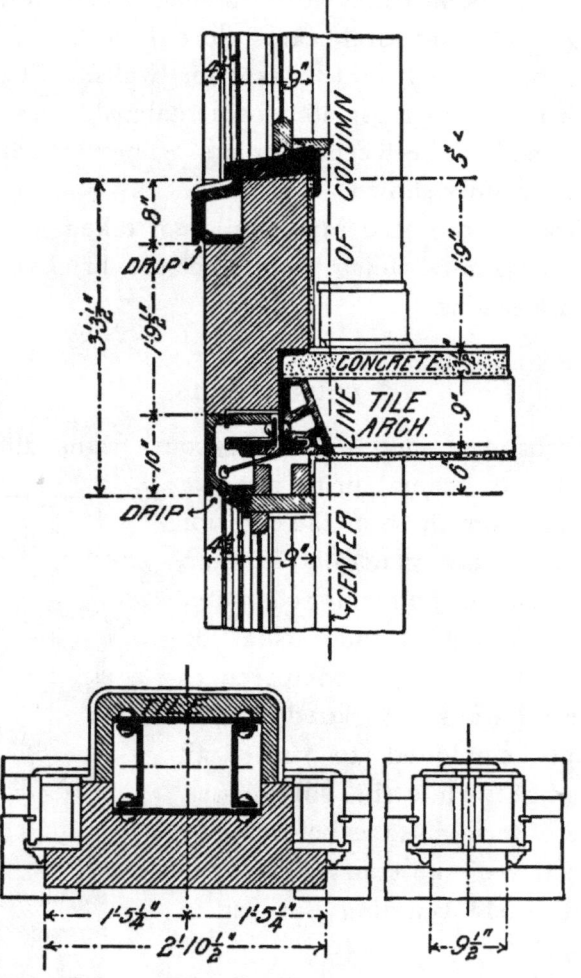

FIG. 62.—Typical Court Wall. Practice of Jenney & Mundie, Architects.

each story must be carried in such a manner as to make it independent of the other stories. This is accomplished by means of brackets at each floor level, and in order that the bracket loads may not become too heavy the bay-window walls must be constructed as light as possible. No yielding or deflection is permissible in these brackets, and if the

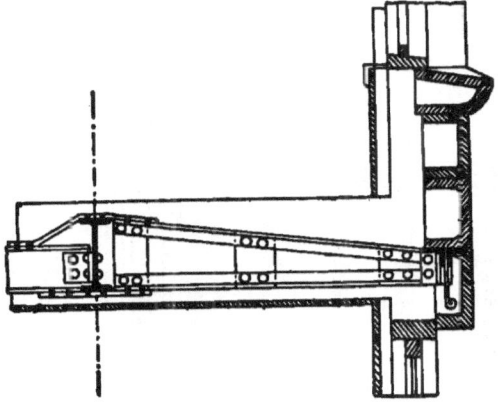

Fig. 63.

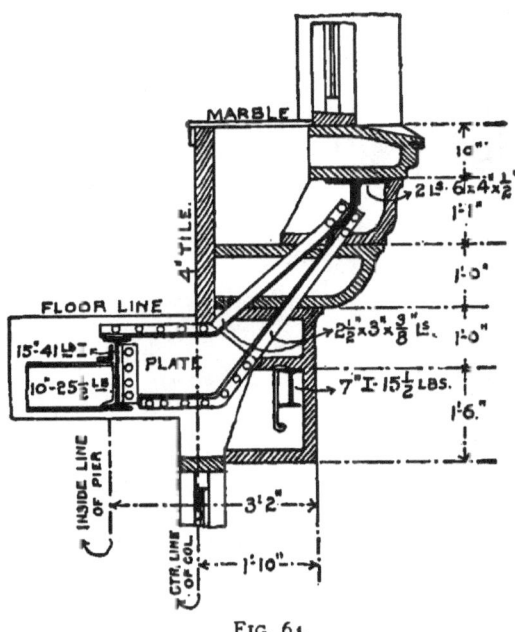

Fig. 64.

supporting member is a floor-beam or floor-girder, as in Fig. 63, taken through a bay window of the Masonic

110   ARCHITECTURAL ENGINEERING.

Temple, the girder should be rigidly connected to the floor system, to prevent any twisting tendency due to the weight of the bay. This is accomplished, as in the above-

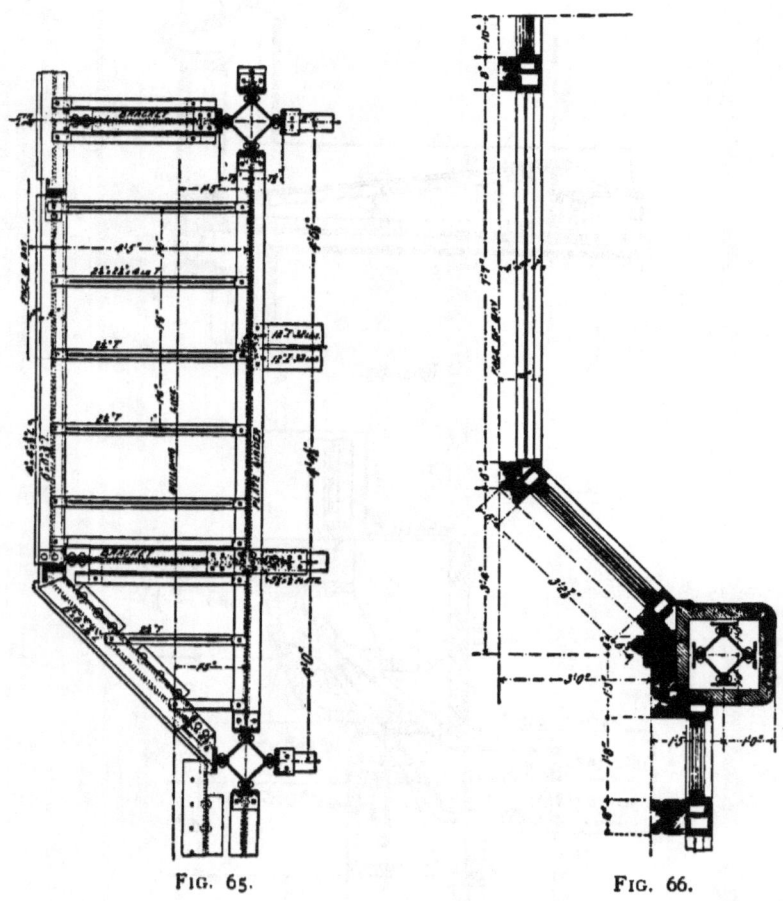

FIG. 65.   FIG. 66.

mentioned figure, by means of the top and bottom tie-plates shown.

Fig. 64 shows a section at the bottom of a bay window in the Masonic Temple.

Fig. 65 shows a half plan of the metal framing for the State Street bay window in the Reliance Building.

SPANDRELS AND SPANDREL SECTIONS.

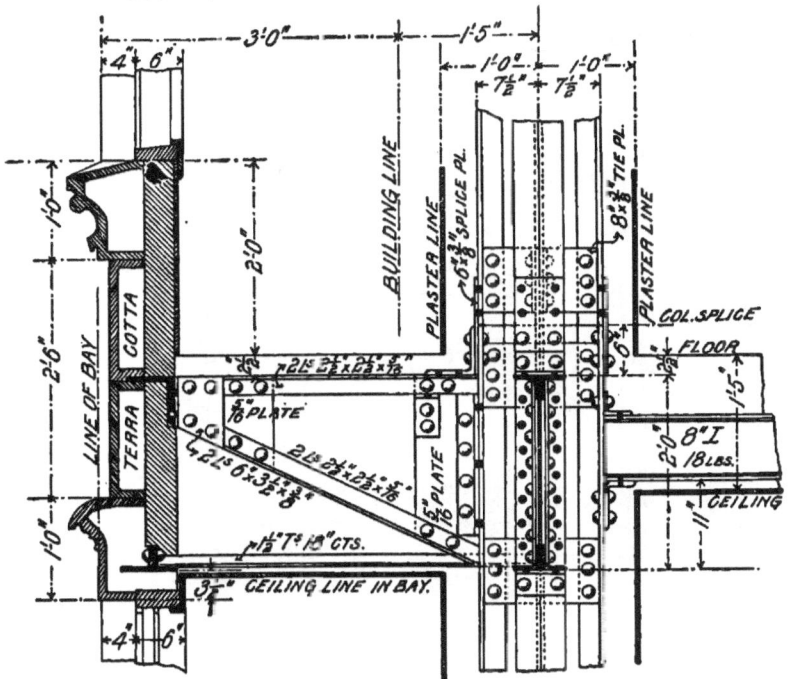

FIG. 67.

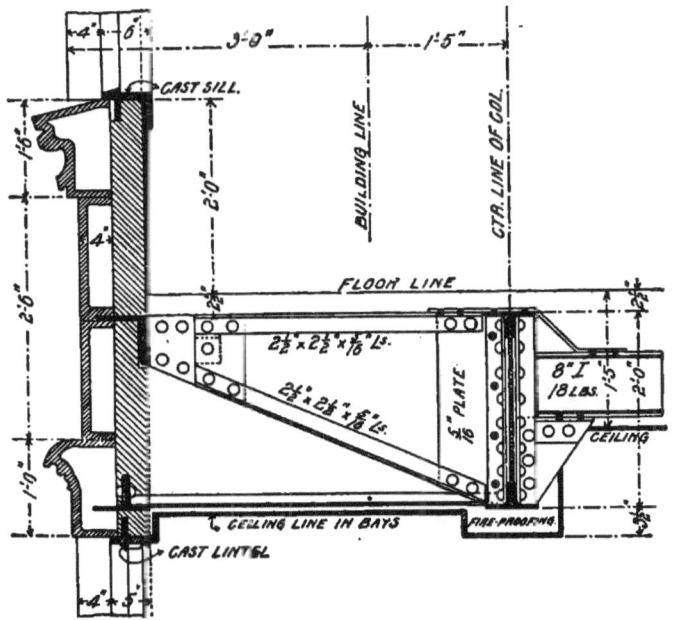

FIG. 68.

The terra-cotta mullions of the bay and the pier are shown in plan in Fig. 66.

The column bracket in the bay is given in Fig. 67, while Fig. 68 is a section at the side bracket.

The method of supporting the floors and ceilings in the bays is shown in Fig. 69.

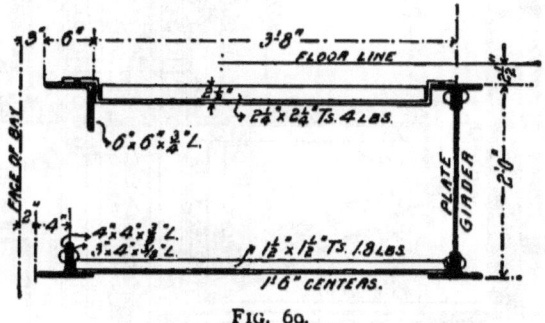

Fig. 69.

CHAPTER VII.

COLUMNS.

THE subject of the interior columns forms one of the most important steps in the modern problem of design, and greater variations are probably to be found here than in any other of the vital features in iron or steel construction. The many forms of columns now in the building market, each having its own enthusiasts, and the many types of connections between the columns themselves and with the floor system, permit of a choice from a dozen or more types, with the details varying widely in each case, to suit the shape chosen. We shall endeavor to investigate the more prominent forms, and point out the advantages and disadvantages of each one. The most satisfactory for general and specific cases may then be selected, as combining the features desired.

A discussion as to the relative values of cast *versus* wrought columns should hardly seem necessary at the present time, but the repeated use of the cast-iron column in ten- to sixteen-storied buildings, and even higher (as shown by their use in the new Manhattan Life Insurance Building of seventeen stories), shows that the questionable economy of cast columns does still, in the opinion of some architects, compensate for the dangers incident to their use. The best practice has declared so uniformly, during the last few years, in favor of the steel columns that the employment of cast metal is new pretty generally confined to buildings of very moderate height or to special cases where advantages are to be gained, as in the use of a number of ornamental cast columns. The great uncertainty as to the uniformity of cast metal led to the use of a very low unit-strain, while in the case of steel the unit-strains can be assumed on a very definite reliance on the trustworthiness of the metal. Among our more progressive designers the use of cast-iron in large buildings has become a thing of the past, and would no more be seriously considered than would the use of cast-iron compression-members in bridges.

Considering the cast sections in more general use as columns, the circular, square, and H-shaped, and their individual connections (see Fig. 70), it will be seen that these splices cannot result in as rigid a framework as the riveted joints in steel-work. The columns in the modern design must be capable of affording stiff connections so as to withstand both the direct dead and live loads transferred from the floor system, as well as sufficient connections for the wind bracing. These cannot be secured well by means of bolts passing through the horizontal flanges of cast columns, even if the workmanship be considered accurate. The workmanship, however, can seldom, if ever, be relied upon as perfect; the bolts never completely fill their holes, and "shims" are constantly employed to plumb the columns. These constitute elements of weakness which may easily allow considerable distortion. The girder connections to the columns, resting on cast brackets, and bolted through the flanges, are bad in the extreme, espe-

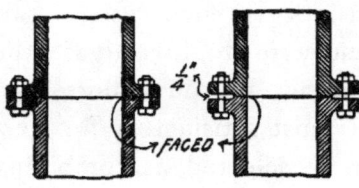
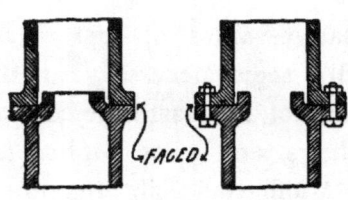

FIG. 70.

cially for cases of eccentric loading and the irregular placing of beams.

To offset these dangers of weak design it is true that cast columns are cheaper per pound and perhaps easier of erection than the steel—considerations that naturally have much weight with the owner of the building. But considering the risks that are run, as in the building at 14 Maiden Lane, New York, which was blown eleven inches out of plumb through the inability of the cast columns to resist the wind pressure, it is hard to understand why architects will persist in the use of such methods, even if requested by the owner. Cast iron, in spite of its apparent stiffness, has a much lower coefficient of elasticity than steel, breaking suddenly when it breaks, while steel suffers distortion.

Steel is now being rolled at such a low price that, considering the extra weight necessary in cast iron, on account of its unreliability, the saving in cost by the use of the latter will be found to be small indeed, even disregarding the dangers assumed by its use.

The more prominent forms of American wrought columns include the Phœnix, Keystone octagonal, latticed angles, channels and lattice, plates and angles, Z-bar columns, and the newer Larimer and Gray types. The relative advantages of these various sections are of the greatest importance, as affecting economical and successful design. In actual practice the treatment of these different shapes will be found to vary greatly with the designer—not only in the relative value of the sections, but in the treatment of any one section. In the first place, the formulæ differ greatly, not in fundamental principles, perhaps, but in the treatment, being often empirical, and containing factors deduced from some special case. These formulæ also generally assume ideal loading, which will seldom occur in the modern building, and no or very few full-sized tests have ever been made on the effects of eccentric loading. Indeed, the full-sized tests on columns of concentric loads even, have been far too limited to show the relative values of the most ordinary column sections.

Burr, in his "Strength and Resistance of Materials," states that "The general principles which govern the resistance of built columns may be summed up as follows:

"The material should be disposed as far as possible from the neutral axis of the cross-section, thereby increasing R;

"There should be no initial internal stress;

"The individual portions of the column should be mutually supporting;

"The individual portions of the column should be so firmly secured to each other that no relative motion can take place, in order that the column may fail as a whole, thus maintaining the original value of R."

The experiments given by Burr would seem to indicate that a closed column is stronger than an open one, due to the fact that the edges of the segments are mutually supporting when held in contact by complete closure. From a theoretical standpoint, therefore, the Phœnix column is undoubtedly the most favorable form for compression, as it forms a closed, and thus mutually supporting, section; and because the capacity of columns of equal areas varies as the metal is removed from the neutral axis. It must also be remembered that any form of column, having a maximum and minimum radius of gyration, is not economical for use under a single concentric load, as the calculations must be based on the minimum radius of gyration. The metal represented by the excess of the maximum radius of gyration is of necessity disregarded, and part of the section is thus lost or wasted, when we consider the ideal efficiency of the column. But practice does not always support theory, and many other questions besides mere form arise in connection with the judicious choice of a section. Indeed, we shall see that several practical considerations in the use of columns in buildings call for a form very different from the ideal circular section; such points as the transfer of loads to the centre of the section, the maximum efficiency under eccentric loading, and the requirements for pipe-space around or included in the column form, all tend seriously to restrict the use of closed or circular sections.

In the column formula, $p = \dfrac{f}{p + a\dfrac{l^2}{r^2} + \dfrac{x_0 x_1}{r^2}}$ (the form of Gordon's formula, including the effect of eccentric loading), there are expressions for the three kinds of stresses in a column under compression—that due to the flexure of the column, that due to eccentric loading, and that due to the uniformly distributed load. The term of eccentric loading does not occur in the so-called Gordon's formula, or in those derived from it, but in building construction this term must not be omitted. The placing of columns centrally over one another necessitates the applications of loads to the sides of the columns, and unless the loads are equal, and on opposite sides of the column, the effect is to increase the stress on the side where the greater load occurs.

The second term in the denominator, $a\dfrac{l^2}{r^2}$, is usually so small that it really makes this term of the least importance in the above equation, due to the ordinarily short length of columns in buildings, and to their usual broad flat bases. Hence in one-story columns (unless in long first-story columns), where the length is usually under 90 radii, the difference in the strength of the various sections tends to disappear, and almost any of the sections will answer with the ordinary unit-strains, if the columns are well made and the loads are not eccentric. Eccentric loading will be considered later, under a general discussion of the various sections.

In longer columns, however, where the length is greater than 90 radii, calculation by the radius of gyration becomes necessary. In the new Schiller Theatre Building, Chicago, Phœnix columns were used, of a length of 92 ft. 10 in., weighing 25,000 lbs. each. Modern building methods have rapidly developed the necessity for columns of extraordinary length, carrying loads hitherto considered visionary. It is not uncommon to have 800 tons and even more on a single column with a sectional area of 158 sq. in. The Edison Electric Illuminating Company of New York City used columns of the Phœnix type, having loads of 600 net tons, 35 ft. 4 in. over all in length, weighing 15,000 lbs. each. As vibration occurred in the building, very low unit-strains were allowed, the columns being further strengthened by disregarding the increment to the least radius of gyration caused by using eight fillers, each $1\frac{3}{8}$ in. thick.

The formula used was one deduced from the experiments at the Watertown Arsenal, namely, $\dfrac{P}{S} = \dfrac{42{,}000}{1 + \left(\dfrac{1}{50{,}000} \times \dfrac{l''}{r^2}\right)}$

for the crushing strain per sq. in.

Twelve-section Phœnix columns were also used in the Chicago Board of Trade, 90 ft. unsupported length, 3 ft. 3 in. diameter, fire-proofed.

But by far the larger number of columns used in modern building construction are, as has before been stated, under 90 radii, being used in single-story lengths of from 10 to 14 feet. The determining factors are, therefore, such practical considerations as affect columns of these lengths; so that the ideal disposition of the metal must be considered in connection with other very important requirements. The following points of the problem are important, in a discussion of which the writer partly follows the points enumerated by Mr. C. T. Purdy, in the *Engineering News*, December 5, 1891:

1. Cost, availability.
2. Shopwork, and workmanship of column.
3. Ability to transfer loads to centre of column—eccentric loading.
4. Convenient connections of floor system.
5. Relation of size of section to small columns.
6. Fire-proofing capabilities of the section.

Points 1 and 2 are of the greatest importance to the owner and builder, and often govern the selection of the column. Points 3, 4, and 5 are for the engineer's consideration, while point 6 is of chief interest to the architect and decorator.

1. *Cost, Availability.*—The question of the cost of the material as it comes from the mill is a purely commercial one, depending upon the market price per pound of the section used. The break in the combine which formerly existed on I beams and channels has reduced the price on these sections from the former combine price of $3.20 to about $1.50, or to a price uniform with that for plates and angles. Indeed the price of iron and steel shapes has never been so low in the history of this country as at the present time, and were such prices to continue, they would doubtless prove a tremendous stimulus to steel construction even in dwellings.

All of the "patent" columns, such as Z-bar, Phœnix, Keystone octagonal, Larimer, and Gray forms, have the great disadvantage of being rolled or manufactured by certain mills only, and in this age of push and hurry the quick delivery of material is a very essential point. The demands for structural steel at good seasons of trade in this country, are so great that it is next to impossible to secure such a prompt delivery of material as is required for the completion of a large building within the contract time. The contracts that have been executed in the city of Chicago during the last three or four years have undoubtedly shown the most wonderful construction in points of excellence and time that the world has ever seen; while it is said of a large building in New York City that the masonry for the twelfth story was laid before the mortar at the first-floor level was dry. The patent on the more important of the patent sections has, however, recently expired, so that now the Z section is being rolled by several mills, and it is not only cheaper than formerly, but much more available, being rolled even on the Pacific coast. The Phœnix shape, although the patent has long since expired, is rolled by but one mill in this country, the Phœnixville, and by one other mill in England. The Keystone column is but little used. Columns of plates and angles, or channels, possess this advantage of availability in a greater measure than any of the other sections, the parts being obtainable at any mill, if not in stock.

2. *Shopwork and Workmanship.*—With the present uniform low price per pound of most of the column sections, the items of shopwork and workmanship become of far greater importance in the cost of the completed column than the cost of the section at the mill—assuming the sectional area, and hence the weight per foot, to be the same. Lattice bars, fillers, gussets, etc., add just so much more weight, without increasing the section, and must therefore be considered from an economical standpoint. The methods of riveting the sections together in the various forms must also be taken into account.

The number of punching operations, as well as the expense of rolling the sections employed, will need to be considered as affecting the cost of shopwork. Thus in the

Gray column no less than sixteen operations of punching are required for four rows of rivets, with the additional expense of hydraulic pressed bent plates, connecting the angles. This will materially increase the cost of manufacture. (See following table.)

Larimer column, 1 row of rivets.

Z-bar column, without covers, 2 rows.

4-section Phœnix column, 4 rows.

Channel column with plates or lattice, 4 rows.

Gray column, 4 rows.

Keystone octagonal column, 4 rows.

Z-bar column, with single covers, 6 rows.

Box column of plates and angles, 8 rows.

Latticed angle column, 8 rows.

8-section Phœnix column, 8 rows.

Z-bar column with double covers, 10 rows.

The new Larimer column, but recently placed on the market by Jones & Laughlins, and first used in Chicago in the Newberry Library building, consists of two I-beam sections bent down along the middle of the web, the two beams being riveted together with a small I-beam filler between. The rivets are spaced 3 in. centres for about 18 in. from each end of the column, and then 5 in. centres.

Where necessary to strengthen the column, this filler is made of two channel-sections, back to back, extending out on either side as far as necessary. Small angles are riveted to the faces of the I beams, and a plate is riveted across the top, on which the girders and column rest (Fig. 71). Where

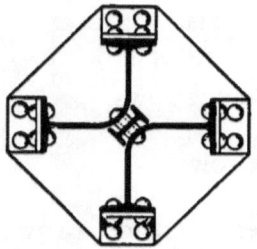
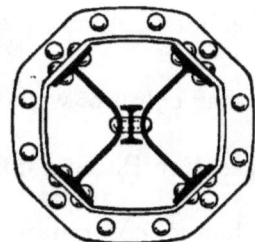

Fig. 71.   Fig. 72.

only two girders occur, the remaining faces are used to rivet the upper column to the plate. Another method has been used instead of the small angles, in the shape of a square or octagonal sheet which is cut from the centre out, part way to the edge, and the lips so formed are bent down in a press, thus making a solid and continuous angle. Still another detail has been made by pressing out in a hydraulic machine a circular sheet to conform in the lower part to the shape of the outside of the flanges of the column (Fig. 72). In this way not only the upper flange, but the vertical flange too, is made continuous around the top of the column. Also the thickness of the horizontal flange is retained uniform, the thickness of the vertical flange being somewhat tapered.

This column is one of the cheapest on the market at the present time, but it possesses one great disadvantage in the smaller sized columns. This lies in the difficulty of driving the rivets that connect the bracket angles with the I-beam flanges. In a 6-in. column, where 5-in. I beams are used, or in smaller columns, it is often very difficult on account of interference to drive the rivets through the holes, unless the rivets are driven in a slanting direction. This often results in weak connections. Jones & Laughlins, the manufacturers of this column, have made a large number of tests of built columns, showing a marked gain in ultimate strength over the Z-bar column tests published by C. L. Strobel. Comparing a 7" Larimer column (6" I beams, 12¾ lbs. per foot, of sectional area of 9.261 □", total length 120", gauged length 100") with a 6" Z-bar column (6" × 3" Z's, ¼" metal, area = 9.32 □", total length = 119.88", gauged length = 80"), the Larimer shows an ultimate strength of 346,300 lbs., or 37,393 lbs. per sq. in., as compared with an ultimate strength of 293,200 lbs., or 31,460 lbs. per sq. in. for the Z column. The Z-bar column failed through the buckling of the Z's, and twisted in a spiral direction between the two ends. The Larimer column deflected in an oblique direction. Larger Larimer columns also showed a greater ultimate strength per sq. in. than the Z-bar columns.

A point that has always been made much of in the claims for the Z-bar column is that but two rows of rivets are required, and those near the centre of the column; for it is reasonable to suppose that punching in the outer portions of a column tends to weaken the member, even when the riveting is most carefully done; and this is even more important in small columns, where the ratio of the radius of gyration to the length of the column is greatest, and where we desire the greatest efficiency of the material used. But is this claim of two rows of rivets founded on fact? If Z-bar columns were used without cover-plates, the claim would indeed be true, but take, for instance, the large Z-bar columns in the Venetian Building, quoted by Mr. Purdy in the article named. No less than ten rows of rivets are required with the heavy cover-plates used, and, indeed, when we stop to consider the large proportion of Z-bar columns which have covers, the claim of only two rows of rivets assumes but little value, and the section proves less desirable than the box column of plates and angles,—inasmuch as the material of the Z's, near the centre of the column, is practically wasted, though adding so materially to the weight. It can hardly be denied, even by the most enthusiastic supporters of the Z section, that the use of this shape has really been thrust upon the Chicago builders during the last few years far more than its merits would warrant. A glance at the Appendix table shows that twenty-two out of a total of forty buildings in Chicago have used the Z-bar column. Its use in Eastern cities has been far more limited.

It is hard to see, therefore, where the Z-bar column possesses any decided advantage so far as shopwork is concerned, unless used without cover-plates. The columns of plates and angles and the Z sections are about on a par in these respects, while the channel columns are more favorable than either. The channel columns are, however, somewhat limited as to section, while plates and angles can be increased to any desired area. The latter section was used in the highest steel building in Chicago, the Masonic Temple, latticing being used on two sides of the columns in the upper stories.

The character of *workmanship* will vary with the different shops, as well as with the different sections used. The reputation of the shop, aided by careful inspection, will determine the excellence of the workmanship.

3. *Ability to Transfer Loads to Centre of Column—Eccentric Loading.*—It will be seen at a glance that many of the sections under consideration are totally unfitted for the transfer of loads to the centre of the column. The conditions in designing a framework are seldom so favorable as not to require many of the columns to be loaded unsym-

metrically, and this point has been carefully considered in the details of the best modern structures, in order to obtain the highest possible efficiency in the material used. Every step in this direction will certainly add to the capacity of the column, for an eccentric load will necessitate the use of a much less mean unit-strain than where the force can be applied directly to the axis. Fig. 73 shows the connection between beams or girders and the Gray column. It is evident that, unless the top of the column is very rigidly bound

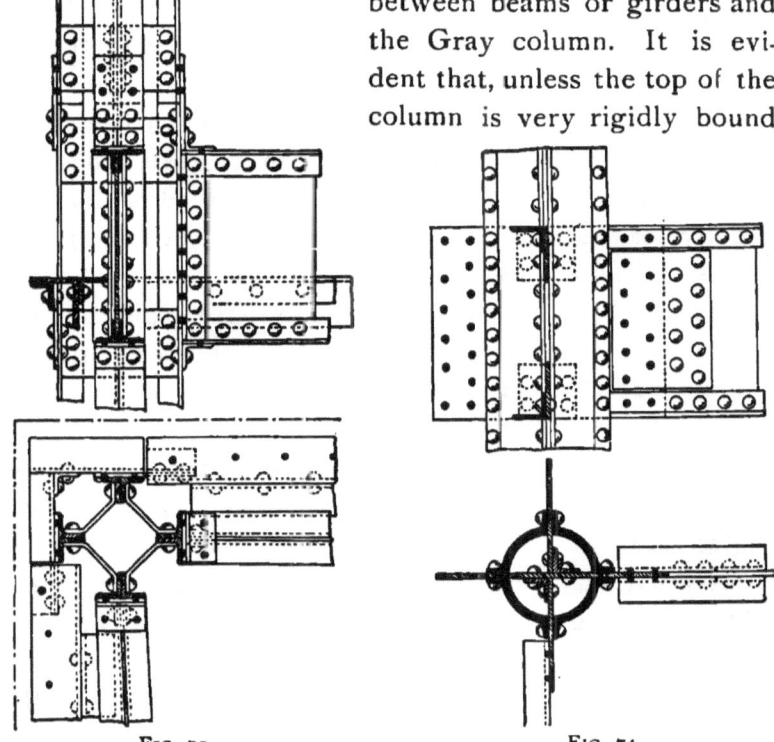

FIG. 73. FIG. 74.

together by outside plates or angles, the girder loads, if eccentric, are borne mainly by the T shape to which the girder is connected, and not by the whole column. This lack of latticing to transmit shear may constitute a very serious disadvantage in cases of heavy eccentric loading.

The use of Phœnix plates with pintle connections, as advocated by Foster Milliken, would certainly seem to possess the greatest advantages under this heading (Fig. 74). There is no leverage in this method to tear the joint asunder, as there is in any flange joint. This system was recently used in the large power-house of the Broadway cable road in New York, with pintle-plates over eight feet deep. Unless pintle-plates can be used, however, any form of closed column is bad under this consideration of central loads, and here the practical method of loading columns conflicts seriously with the use of an ideal closed section.

The Z-bar column possesses advantages here, too, over most of the forms of closed columns, and even when coverplates are used this is so (though not in as great a degree), as the column may almost always be turned so that the heavily loaded beam may be introduced between the Z flanges. This advantage is especially great at the tops of buildings where small columns without cover-plates carry beams with heavy loads, for here the column is open on all four sides, so that all loads may be taken to the centre of the column. (But Z-bar columns without covers fail by wrinkling, and under this condition they are the weakest of any of the sections.) The box column of plates and angles, however, possesses this same advantage, though not to as great an extent in the lighter sections. The possibility of changing the section of a column so that the radius of gyration shall be greater or less in either direction across the section must not be overlooked, for if all the loads occur on one side of a column, it is a great advantage to have the radius of gyration greater in the line of the load.

The calculation for eccentric loading should be treated a follows:

(*a*) Determine the section required for the *total* load, both eccentric and concentric, the whole considered as concentric.

(b) Find y_1, or half the width of the column.

(c) Find the radius of gyration in the plane of eccentric loading.

(d) Find the area of section required to resist the bending moment arising from the eccentric loading, using radius of gyration and y_1 as in the assumed section. The moment due to eccentric loading will equal the eccentric load × its distance of application from the axis of column, or

$$M_0 = \frac{fI}{y_1} = \frac{fAr^2}{y_1}, \text{ whence } A = \frac{M_0 y_1}{fr^2}.$$

(e) If this second area can be added to the first assumed area of section without changing the radius of gyration and y_1 materially, it may be done, thus obtaining the total area of section without a new solution.

(f) If, however, the radius of gyration and y_1 are changed materially, in providing for the new area required, then a new assumed sectional area is taken, radius of gyration and y_1 found for it, the solution proceeding as before.

4. *Convenient Connections.*—This feature in column construction is a very important one. Satisfactory details can easily be made for almost any of the sections, where the beams are symmetrically placed and loaded, and where all occur at the same elevation; but when the irregular placing of beams is necessitated, as regards position, load, and height, it is important that the character of the column afford as great an opportunity as possible for the connection of

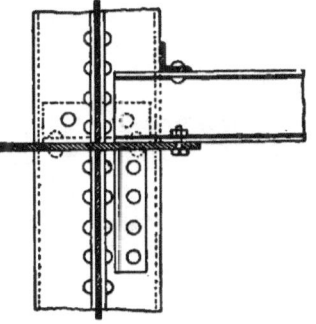

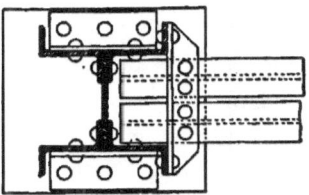

Fig. 75.

plates and angles. The connection in Z-bar columns forms one of the greatest advantages in the use of this section; and in the smaller columns without covers, where the connections are generally the most difficult, the advantages are the greatest. The general system of connections is shown in Fig. 75, taken from the Monadnock Building.

Angle-brackets are riveted to the column, on which is placed a plate $\frac{1}{2}$ in. to 1 in. in thickness, on top of which come the girders, the column of the next floor setting centrally over the one below. The girders are riveted or bolted through to the bed-plate below, by the flanges, and through an angle above, as shown in Fig. 75. A small wrought-iron "gib" or wedge is dropped in between the top end of the girder and the web, to take up any possible compressive strains. If the girders are all to be brought to one level, cast-iron bolsters are used.

The system followed in the Phœnix column is as shown in Fig. 76, consisting of angles riveted to the extended

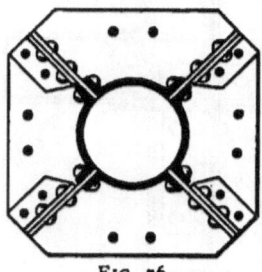

FIG. 76.

fillers, on which a plate is placed, holding the girders and the superimposed column. The upper column is held down by angles riveted to the bed-plate. Under eccentric loading a considerable tilting movement occurs in this column, unless used with pintle-plates, as before suggested. Connections were made with bent plates in the Old Colony Building, Chicago, as shown in Fig. 77.

Box columns of plates and angles offer quite as many advantages as regards connections, if not more, than any other section. The details are really the simplest of all, when we consider columns of a single floor height only (Fig. 78), but the joint is not a desirable one, nor is any where a horizontal plate separates the two columns; for it prevents efficient splicing, as well as good girder-connections. This point will be taken up later under the head of "Column Joints."

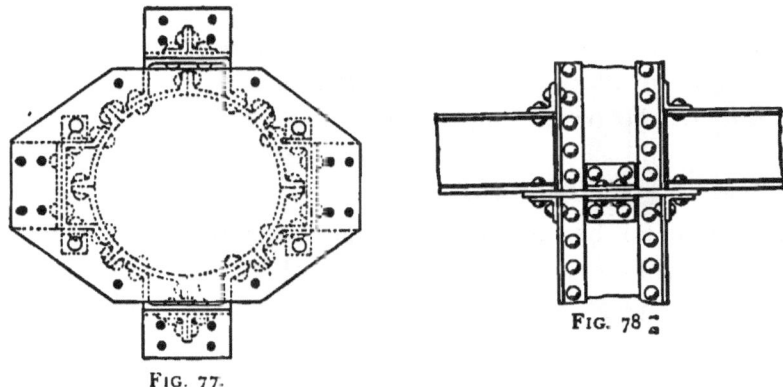

Fig. 77.

Fig. 78.

5. *Relation of Size of Section to Small Columns.*—It is not generally desirable in building construction to have a very small column in the upper stories, because girder loads are so much heavier, proportionately, than the column loads. Sometimes as many as six beams must connect with an upper-story column at one level, and in such cases it is almost impossible to make good connections with a small column.

6. *Fire-proofing Capabilities of the Section.*—The rectangular column sections will not, of course, fire-proof as compactly as the circular sections, but when the room thus lost is used for "pipe-space," as is becoming more and more frequent, this point has great value in the estimation of architects. In the Columbus Building, Chicago (1893), a square hole was cut in all of the bed-plates of the columns to allow the passage of pipes inside of the column area. Such a cutting of bed-plates cannot be too severely condemned. The increased use, however, of vertical splices in columns, instead of horizontal bed- and cap-plates, allows all water-, waste-, and vent-pipes to be carried up along the side of the metal columns, and inside the fire-proofing slabs,

where the room may be had without too much waste. It is not advisable to place any piping inside of the metal columns, and hence such sections as the Phœnix and Keystone-octagonal offer no advantages in this respect. The columns of plates and angles, channels, Z's and the Gray column, all allow considerable pipe-space within the minimum circular or rectangular enclosure for fire-proofing.

It would seem, however, that separate ducts in the walls, or along the sides of the columns for all piping would be far better than such concealed risers. Separate ducts would result in increased outlay, but they would offer the great advantage of allowing inspection of all piping whenever and wherever desired.

The largest Z-column section in "The Fair" building, Chicago, consists of 4 Z bars $6'' \times \frac{7}{8}''$, 2 webs $16'' \times \frac{3}{4}''$, 6 covers $16'' \times 1\frac{3}{8}''$, aggregating an area of 142 sq. in. and carrying a load of 1,700,000 lbs. The largest Z column in the new Y. M. C. A. Building, Chicago (see Fig. 79), was a two-story column 24' 3" long, composed as follows: 4 Z's $6'' \times 3'' \times \frac{7}{8}''$, 2 plates $24'' \times \frac{7}{8}''$, 2 plates $16'' \times \frac{7}{8}''$, 1 plate $14'' \times \frac{3}{4}''$, 2 plates $26'' \times \frac{7}{8}''$, 4 angles $4'' \times 4'' \times 1\frac{3}{8}''$, 4 angles $5'' \times 4'' \times \frac{7}{8}''$ — total = 218 sq. in. The minimum Z section generally used is 4 Z's $3'' \times \frac{5}{16}''$, 1 web $8'' \times \frac{5}{16}'' = 12.4$ sq. in. Metal less than $\frac{5}{16}''$ in thickness is never used in the best practice.

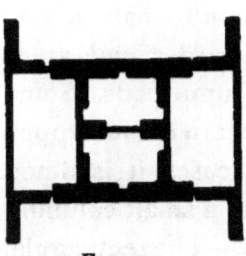

FIG. 79.

The calculations of the strengths of wrought columns, in accordance with the building laws of New York, Boston, and Chicago, are given in Chapter XII; and the unit-strains used in several prominent buildings are given in Chapter XI.

It is apparent, therefore, that each of the types of

columns considered, has its own good points, but the choice of *one*, as decidedly superior to all others, would be wellnigh impossible. The Larimer column may lead in cheapness, the Z or box columns are superior for connections, the material in the Phœnix or Keystone columns is placed most advantageously from a theoretical standpoint. The choice, then, must depend on the personal views of the designer, as well as on the local conditions as to cost, manufacture, and the details employed in the problem at hand. The writer favors the box column of plates and angles. It is easily obtained, cheap, good for connections, possesses a minimum and maximum radius of gyration, which can be utilized under eccentric loading, and it offers the greatest advantages for continuous columns, a point which will be considered later in connection with wind bracing.

A discussion on columns would hardly be complete without some reference to the views expressed on this subject by Gen. Wm. Sooysmith. He advises the use of limestone pillars instead of steel columns, declaring that the action of the metal-work under heat would be dangerous in the extreme. To quote: "There *may* be steel buildings in which the fire-proofing has been so well done that they will pass through an ordinary fire without such failure. But if the steel becomes even moderately heated, its stiffness will be measurably diminished, and the strength of the upright members so reduced as to cause them to bend and yield." While acknowledging the great experience and ability of Gen. Sooysmith in constructive work, and especially in foundations, the writer would seriously question the authority for such an apparent reflection on fire-proofing methods. There not only *may* be buildings which are sufficiently fire-proofed, but it is a well-established fact that builders, architects, and engineers can and do fire-proof their buildings sufficiently to guard against all possible heat arising from the material used in the building, or from the burning of surrounding structures. And that there is almost no limit to the possibility of protection from heat by fire-clay is shown in the immense converters in use by the large steel companies. They are made of steel, protected by fire-clay, and in spite of a temperature of 2000° night and day, these furnaces last even as long as four years before renewal.

Again, limestone ($CaCO_3$) is friable under the action of heat, decomposing into lime (CaO) and carbon di-oxide (CO_2) at a temperature of 600°. Hence the limestone pillars would require quite as much protection by fire-proofing as the steelwork. Gen. Sooysmith claims a safe load of 500 tons for a column of limestone $2' \times 2'$ in area and $9'$ high. This equals 576 sq. in., or at 5500 lbs. per sq. in. given by Rankine, gives an ultimate compressive resistance of 1584 tons. Allowing the factor of safety of 8, recommended by Rankine, we have even less than 200 tons, while Baker recommends but 20 or 25 tons per sq. ft. for the best ashlar masonry (10 tons was the maximum pressure in the Brooklyn bridge, and 19 tons in the St. Louis bridge), or 100 tons for this limestone column. This same load of 200 tons would be carried by a 12″ Z-bar column of 4 Z's $3'' \times 6'' \times \frac{3}{4}''$, and 1 plate $8'' \times \frac{3}{4}'' = 42$ sq. in. area, at 10,000 lbs. per sq. in. The economy of space in this latter column is at once apparent, even disregarding the fire-proofing necessary to a limestone pillar.

THE FIRE-PROOFING OF COLUMNS.

As the columns carry the greatest loads found in modern buildings (some over 1,500,000 lbs.), the proper fire-proofing of these members becomes a most important subject for consideration. In only too many cases, however, is this slighted even to a very dangerous extent, as was proven by the Athletic Club Building fire, before referred to.

The first attempts at making fire-proof columns were through the use of a double column, one inside the other, with the intervening space filled with plaster. This idea was patented, and reference may still be found to such construction in the New York building laws, as: "The said column or columns shall be either constructed double, that is, an outer and an inner column, the inner alone to be of sufficient strength to sustain safely the weight to be imposed thereon."

The scientific fire-proofing of columns by means of terra-cotta was started by Mr. P. B. Wight in 1874, and the Chicago Club house, designed by Treat & Foltz, architects, was the first instance where terra-cotta gores were used around columns. Many systems have since been introduced, and both the hard tile and the porous tile have been used extensively. The cheapest method has been through the use of shells of hard terra-cotta surrounding the column, but not fastened to the metal-work. This system is decidedly faulty in placing so much reliance in the joints alone for stability, as the blocks are simply hooked to one another, and not to the metal column.

The requirements in the adequate fire-proofing of columns are:

1. The material must be indestructible by fire.
2. The material must be non-heat-conducting.
3. The material must be so secured to the column that it cannot be dislodged.

The use of hard fire-clay tiles is only to be recommended when such tiles are hollow, with a proper air-space around the metal column, and even then experience seems to show that the hard tile is in no way as satisfactory under great heat as the more porous kinds. Applications of cold water in combination with heat have also proved the hard tile far less reliable in case of conflagration than the porous tile. The hard tile is very apt to crack off under such conditions, as has been stated in the chapter on Floors.

The use of solid blocks of porous tile, well bedded against the metal column, seems to be the one most highly recommended. Here, as in terra-cotta floor arches, the competition in price, which places the better article or method at a disadvantage, is to be deplored. Loosely drawn specifications are also responsible in a great measure for many very common defects. All wiring of the individual blocks either to the columns or to one another

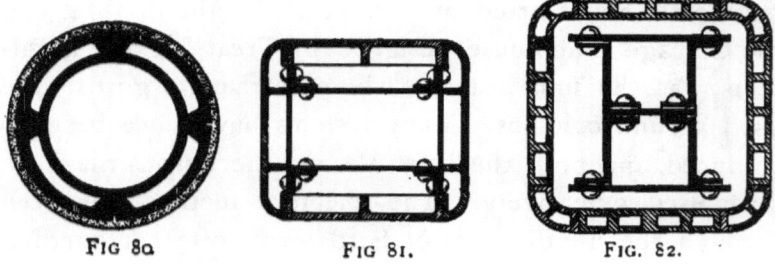

Fig 80.   Fig 81.   Fig. 82.

should be made by means of copper wire. Figs. 80, 81, and 82 show the ordinary methods of placing the fire-proof furring for columns.

The Z-bar columns in the newer portion of the Monadnock Building were fire-proofed as shown in Fig. 83 up to

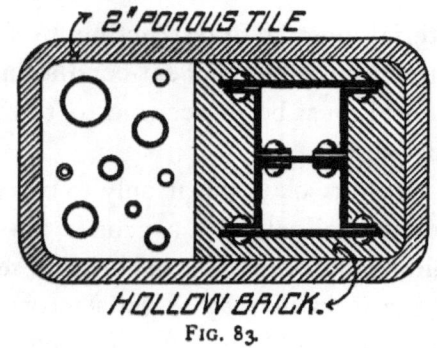

Fig. 83.

and including the eighth floor. Hollow bricks, laid in cement mortar, were built solidly around the columns to a line distant 4 in. from the extreme points of the metal-work, and a 2-in. coating of hollow tile was then laid against the brick backing extending beyond the column in one direction, to serve as a space for vertical pipes. The columns above the eighth floor received the hollow-tile protection only.

The requirements for fire-proofing the interior columns of office buildings are thus defined by the Chicago ordinance:

"The coverings for columns shall be, if of brick, not less than 8 inches thick; if of hollow tile, one covering at least $2\frac{1}{2}$ inches thick. If the fire-proof covering is made of porous terra-cotta, it shall be at least 2 inches thick. Whether hollow tile or porous terra-cotta is used, the courses shall be so anchored and bonded together as to form an independent and stable structure."

"In all cases there shall be on the outside of the tiles a covering of plastering with Portland cement or of other mortar of equal hardness and efficiency when set."

"If plastering on metallic laths be used as fire-proofing for columns, it shall be in two layers, of which the first shall be applied in such manner that the mortar will cover the entire external surface of the column, while the space between the two layers shall be not less than 1 in. thick."

"The metallic lath shall in each case be fastened to metallic furrings, and the plastering upon the same shall be made with cement. Protection for the lower five feet shall be required in this case the same as where porous terra-cotta or hollow-tile covering is used."

CHAPTER VIII.

WIND BRACING.

A CAREFUL comparison of the treatments of wind forces as applied to the mercantile buildings of to-day leads, one to the conclusion that the designers differ very materially in regard to the forces to be resisted, the strength of the materials employed, and the most efficient details of construction. Indeed, there are buildings from ten to sixteen stories high in the city of Chicago, that possess absolutely no metallic sway-bracing, and others, scarcely better, where sway-rods, as wind-laterals, were attached to pins through lugs on the cast columns, which lugs were of an ultimate strength of, perhaps, 25 per cent of the rods. H. H. Quimby, in his paper on "Wind Bracing in High Buildings,"* mentions the case of an office building recently erected, of seventeen stories, or 200 ft. in height, and 60 ft. wide; 13-in. walls were used front and back, broken by windows and bay windows, with wind bracing consisting solely of the interior partitions of 8-in. box tile, with four ribs of $\frac{9}{16}$ in. each, or $2\frac{1}{4}$ in. thickness of tile in each partition. This building towers above its neighbors of five or six stories only, while but a few blocks away is one of seventeen stories, also, but 150 ft. wide, or $2\frac{1}{2}$ times the width of the former, with sway-bracing consisting of 15-in. channel-struts and 6-in. eye-bars. Such is the diversity of practice.

Some architects depend solely upon the partitions of hollow tiles for the lateral stability of their buildings,

---

*Trans. A. S. C. E., Vol. XXVII, No. 3.

weak as the partitions must be through the introduction of numerous doors and office lights. This method of filling in the rectangles of the frame by light partitions *may* be efficient wind bracing, but the best practice would certainly indicate that it cannot be relied upon, or even vaguely estimated.

A building with a well-constructed iron frame should be safe if provided with brick partitions, and if the base is a large proportion of, or equal to the height, or if the exterior of the iron framework is covered with well-built masonry walls of sufficient thickness; for the rigidity of solid walls would exceed that of a braced frame to such an extent that, were the building to sway sufficiently to bring the bracing-rods into play, the walls would be damaged before the rods could be brought into action.

Hence the stability must depend entirely either on the masonry or on the iron framing; and in veneer buildings, which are being considered here in particular, the latter system of bracing the metal-work must be used, with the walls as light as possible, simply enclosing the building against climatic and injurious forces. This practice has been adopted quite uniformly by the best Chicago architects and engineers, and will alone be considered here as a method of wind bracing.

Each building offers its own peculiar conditions to the carrying out of proper wind bracing, and many factors must be considered for a judicious solution. The height, width, shape, and exposure of the structure, as well as the character of the enclosing walls, will determine the amount of the wind pressure to be cared for, while the details of construction, the internal appearance, and the planning of the various floors will largely influence the manner in which this bracing is to be treated. The architectural planning of the offices, rooms, and corridors, often raises most serious obstacles to a proper arrangement of wind bracing, and the engineer is frequently called upon to make most generous concessions in favor of doors, windows, passages, and even whole areas, as is sometimes demanded in banking- or assembly-rooms and the like. Such considerations have led to the development of the portal type of wind bracing. As more and more of the constructional work of large buildings is placed in the care of the engineer, as opposed to the purely architectural or decorative draughtsman, just so will the former insist that a proper regard for construction is of equal value with the artistic portion of the work. The one must supplement the other, instead of giving way to irrationalities of design.

Two distinct corps of workmen are found in the offices of the more prominent architects of the day: the architectural draughtsmen, for all decorative design and work, and the engineers, who have charge of the constructional problems, as indicated in this outline. In such an office the two kinds of work can be carried on simultaneously, concessions made on both sides, and a satisfactory medium reached.

Quimby, in his article on wind bracing, favors the provision of a 40-lb. wind pressure, with iron or steel bracing strained not over $\frac{1}{5}$ of the ultimate strength; while George A. Just, in a discussion, advocates the use of 30 lbs. Circumstances must to a great extent govern the choice of the designer. The shape and exposure of the structure, and the solidity of the enveloping walls, will, as said above, largely determine the amount of wind pressure to be carried by the metallic bracing; but if such bracing be relied upon entirely, a unit of 30 lbs. should serve as a minimum. The following was adopted by E. C. Shankland, Chief Engineer of the World's Columbian Exposition: For roof trusses, 40 lbs. per horizontal sq. ft. of roof taken vertical, or 25 lbs. per sq. ft. taken vertical in addition to the effect of 30 lbs. wind acting under an angle of 20° with the horizon, whichever will give the largest result. On purlins and jack rafters take 30 lbs. per horizontal sq. ft.; on gallery floors take 80 lbs. per horizontal sq. ft.; on main floors take 100 lbs per horizontal sq. ft. A horizontal wind pressure of 30 lbs. per sq. ft. shall be taken care of unless otherwise decided by the Engineer of Construction. All details must be carefully calculated both for bearing and shear.

Many and many are the architects who have used cast-iron columns piled story on story, with tile partitions only as a wind-resisting medium, and their structures stand, to become a source of wonder to the engineering profession. But in a field of such great uncertainty any judicious increase in safety is in the nature of insurance, and must not be regarded as wasted, simply because never destroyed.

Wind bracing must reach to some solid connection at the ground. It should also be arranged in some symmetrical relation to the building outlines. If the building is narrow and braced crosswise with one system, the bracing should be midway, while if two systems are employed,

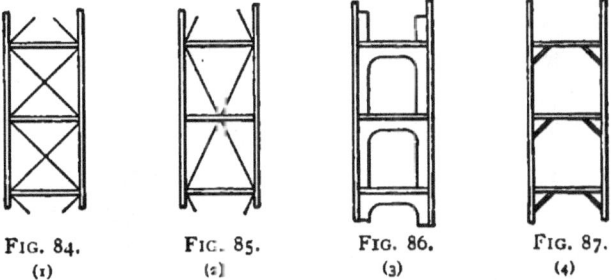

FIG. 84. (1)   FIG. 85. (2)   FIG. 86. (3)   FIG. 87. (4)

they should be placed equidistant from the ends. This symmetry is necessary to secure the equal services of both systems, thus preventing any twisting tendencies.

The more common forms in ordinary practice are shown in Figs. 84, 85, 86, and 87.

Each type must be figured properly, as the strains in the horizontal members and the columns are essentially concerned in the calculations. The problem is not capable of exact solution, owing to several indeterminable factors that enter into the computations, and the consequent equal number of assumptions that must be made. The stresses in the wind bracing will be maximum when the direction of the wind is normal to the exterior wall, or parallel to the plane of bracing. This condition is, therefore, assumed. A further assumption is made that the floors are sufficiently rigid to transmit the horizontal shears due to wind.

The external forces will be the same whichever of the four methods, shown in the figures above, is used, provided the exposed areas, panels, etc., are the same. The horizontal external force at any panel point will be equal to the distance between the systems (at right angles to the bracing) times the distance between floors half-way above and half-way below, times the assumed wind pressure per sq. ft. The total shear at any point equals Σ forces at or above the point taken.

These shears are undoubtedly reduced to some considerable extent through many practical considerations. The dead weight of the structure itself, the resistance to lateral strains offered in the stiff riveted connections between the floor systems and the columns, the stiffening effects of partitions (if continuously and strongly built), and linings, coverings, etc., all tend to decrease the distorting effects of the wind pressure. But, in view of the uncertainty in regard to the efficiency of these latter considerations, they may not be relied upon, and are therefore disregarded in the calculations.

SWAY-RODS (1).

The simplest case of wind bracing is shown in Fig. 84. Considering one bay alone as braced, the system may be analyzed as follows: Referring to the upper story of a framework, as shown in Fig. 88, $P_1 = pH_1L_1$, where P_1 = resultant wind pressure on upper story, p = unit-pressure, and H_1 and L_1 equal respectively the height and width of the area affecting the bracing in the panel under consideration. $\frac{P_1}{2}$ must then be the horizontal component of

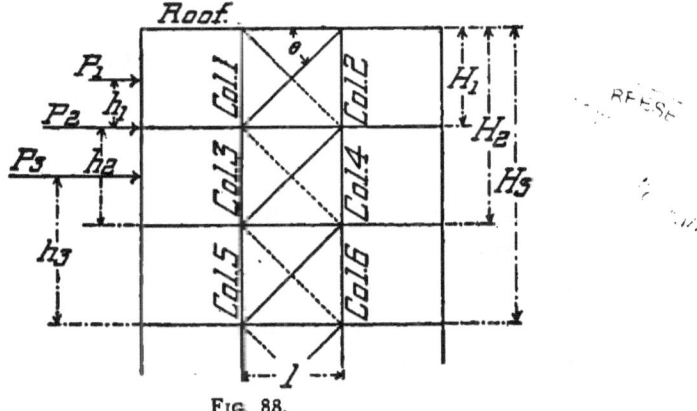

Fig. 88.

the stress in the diagonal, and the tension in this diagonal, making an angle θ with the horizontal, must be

$$T_1 = \frac{P_1}{2} \sec \theta.$$

The diagonal tension in the second story from the top will be $T_2 = \left(\frac{P_1}{2} + P\right) \sec \theta$, where P = wind pressure on any single story, assuming them to be of equal height. $\frac{P_1}{2} + P$ = compressive stress in the horizontal strut at the top-floor level. In like manner, $T_3 = \left(\frac{P_1}{2} + 2P\right) \sec \theta.$

The tension in the diagonal rods will cause a decrease

in loads on the windward columns, and an equal increase in loads on the leeward columns. Calling this increase or decrease V_1, we have

$$V_1 = \frac{P_1 h_1}{l}, \text{ where } h_1 = \frac{H_1}{2}.$$

In a similar manner,

$$V_2 = \frac{P_2 h_2}{l}, \qquad V_3 = \frac{P_3 h_3}{l}.$$

V_2 must equal $V_1 +$ the vertical component of the diagonal T_2, or $V_2 = \frac{P_2 h_2}{l} + T_2 \sin \theta.$ This will serve as a check on the calculations.

These wind loads V_1, V_2 etc., must be added to all the other regular loads on the columns. In the columns 1, 3, etc., the direct or dead loads carried by the columns resist the upward vertical components of the stresses in the rods connected to the bottoms of these columns. Thus the dead load in column 3 is reduced by the full amount of the upward compressive strain from wind in that column, or V_2, and if this amount were to equal or exceed the dead load in column 3, tension would occur in the connection of this column to the one below.

It will be seen that the increment to the stress V at each floor may be eccentric, as shown in Fig. 95, the length of the arm equalling the distance from the point of attachment to the horizontal strut, to the centre of the column itself. If this connection were at the axis of the column, the eccentricity would be reduced to 0, and the eccentric load become a dead load.

Take the case of a typical skeleton building, fourteen stories in height, of 12 feet each, 24 foot front, and columns spaced, 12 feet apart in the depth of the building. Assuming that stiffness against side-yielding alone is necessary, place diagonal members in each story, as in Fig. 89, utilizing the floor-girders as struts, with the columns as chords. At 30 lbs. per square foot wind pressure the panel load equals 4,300 lbs. Considering the protection afforded by neighboring buildings, the point of application of the resultant wind pressure will be taken at two thirds of the height of the structure above ground. The total shear will then equal about 60,000 lbs., or 30 tons. In the basement panel, then, sec $\theta =$ 1.12, giving 33.6 tons tension in the cellar diagonal. The moment of the resultant wind pressure $= 30 \times 118 = 3,540$ foot-tons, and this, divided by 24, gives $147\frac{1}{2}$ tons tension at the windward foundation. The vertical component of the basement diagonal $= 16.8$ tons, leaving a compression of about 131 tons on the leeward column.

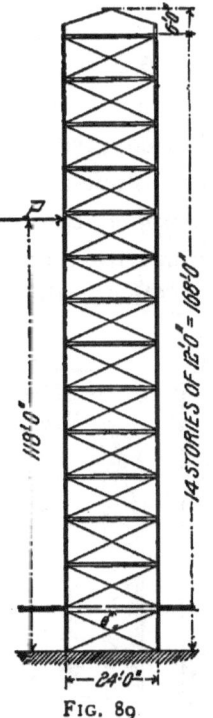

FIG. 89

The dead weight, including iron, walls, floors, filling, etc., will equal about 250 tons for one foundation, while even for a building with no filling or partitions completed, the dead weight is still some 200 tons, thus rendering anchorage unnecessary.

If, in the same cross-section of the building, n bays not adjacent are braced by means of diagonal rods, the tension T becomes $T_1 = \dfrac{P_1}{2n} \sec \theta$, and $V_1 = \dfrac{P_1 h_1}{nl}$.

The bracing in Fig. 85 may easily be analyzed in a manner similar to the above.

The highest building in the city of Chicago is the Masonic Temple, 273' 10" from grade to top of coping. A cross-section of this building is shown in Fig. 90, with one system of bracing-rods. It will be seen that a combination of forms (1) and (2) was used, the bracing being arranged to suit halls and doorways. In this building the sway-rods were not connected to the floor-beams themselves, but to special I beams placed between the columns and just below the floor system.

In "The Fair" Building, system 1 was used, but with lattice girders from column to column, serving as struts and floor-beams at the same time. Gusset-plates were dropped below the girder to receive the pins for connection with the turnbuckle rods.

One of the best examples of system (1) of wind bracing in Chicago, the writer found to be that described by Mr. C. T. Purdy in his article in the *Engineering News* of

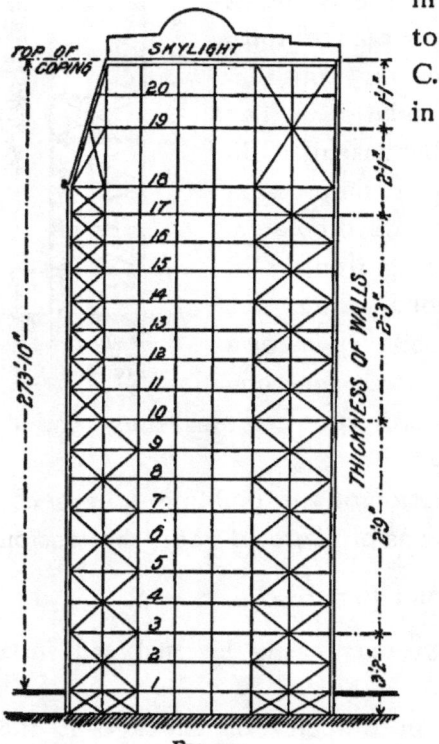

FIG. 90.

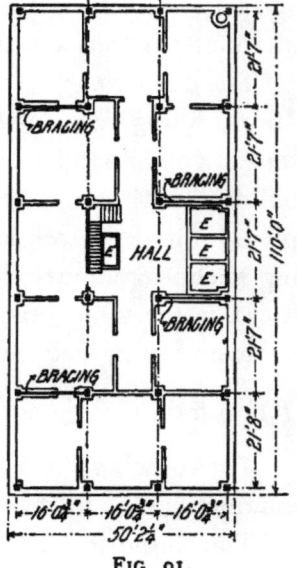

FIG. 91.

December, 1891. This building, the Venetian, is of the veneer type, and contains some excellent details. The floor plan is shown in the accompanying figure (91) with the

four sets of sway-rods given. Each set of bracing is therefore figured to resist a wind pressure for an area, the horizontal width of which is equal to one fifth the depth of the building, and the height of which is the height of the building. The area tributary to each floor × 40 lbs. equals the horizontal shear at each floor or panel-point, while the total shear at any floor equals the sum of the shears acting on the panel-points directly above, as we have seen before. It was not considered necessary, however, to carry the whole amount of this shear into the steel bracing. The practical considerations which tend to diminish the distorting effect due to a lateral force, decided that but 70 per cent of these shears needed to be cared for by the bracing, leaving 30 per cent to be taken up by the other factors. The strains and sections for one bay are here given (Fig. 92).

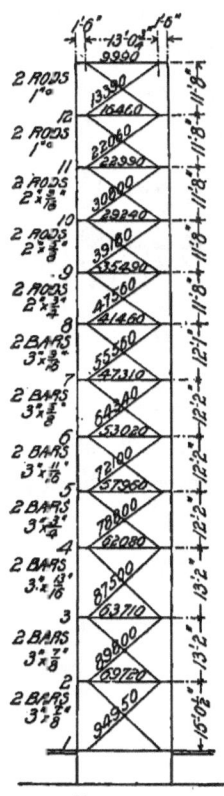

FIG. 92.

All the columns affected by this bracing were made continuous from the foundations to the second-floor level, and portals were used to take the place of the diagonal rods in two instances where rods were out of the question. This occurred on a main floor devoted to large banking-rooms. The bending moments due to these portals were taken up in the columns. In the case where the rods came down to the first-floor level, the bottom strut was connected to the columns so as to take both tension and compression horizontally, as well as to resist the component of the rod strains. This insured the resistance of both columns to the horizontal thrust of the strut, whichever pair of rods was strained, and the columns were calculated to resist the bending moment incurred, as well as to carry the regular column loads.

With the use of the portals, the columns were designed

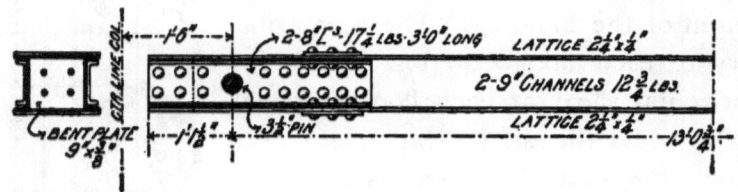

FIG. 93.

to resist the bending moment which the stopping of the rods necessitated, and as a further assurance that these connections should be as strong as the rest of the system, the

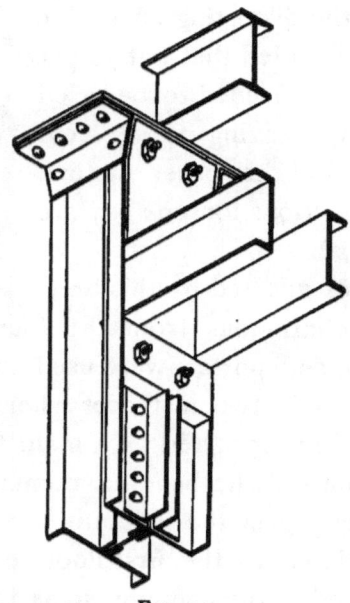

FIG. 94.

top connections of all of the first-floor beams were omitted, and the clearance spaces between all the beams and columns were driven tight with thin metal wedges, until the girders and beams passing along the column axes were continuous and in compression out to the sidewalk walls, which latter are backed by the solid street.

The horizontal channel-struts are shown in Fig. 93. They were used as shown up to and including the seventh floor. A lighter section was used for the floors above. A slight connection only was made between the channel-struts and the columns. The struts were planed at both ends, with no clearance, thus making butt joints with the columns. A bent plate between the channels provided holes for four rivets connecting to the columns, but they were hardly necessary. Underneath the ends of these struts a cast-iron block was bolted to the column and supported by two bracket-angles beneath, with sufficient rivets to resist the vertical compression of the rods in this direction (see Fig. 94).

Above the ends of the struts other cast-iron blocks were used, planed top and bottom, thus allowing them to fit in tightly between the tops of the struts and the cap-plates of the columns. These blocks, therefore, fitted into the recesses made by the flanges of the Z-bars so closely that the $\frac{3}{4}''$ cap-plates were brought into direct shear entirely around three sides of the blocks. The shear resistance of the plate, together with the weight of the beam on it, was more than sufficient to resist the upward vertical component of the rods. Such cast-iron blocks in this connection are very convenient for use, for it often happens that the bracket-angles cannot be brought directly

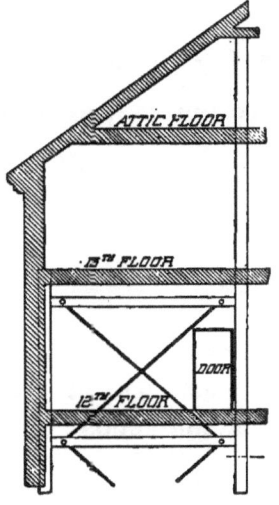

FIG. 95.

under the channels of the struts, and the medium between the strut and the bracket-angles must act as a beam as well as a filler. Fig. 95 shows a partial cross-section of the building with doorway, etc. This shows the reason for placing the pin-points so far from the column centres. The channel-struts are reinforced with cover-channels to resist the bending moment on the strut caused by thus moving the pin-centres.

The diagonal rods in this building were proportioned on a basis of 20,000 lbs. per sq. in. All rods had turnbuckles, and no rods were of an area less than $\frac{7}{8}''$ square. The Ashland Block, by Burnham & Root, Chicago, has longer struts than those in the Venetian Building, 15'' channels being used in the floors, acting both as struts and floor-beams.

PORTAL BRACING (3).

The third method of wind bracing, called the portal system, may be analyzed as follows (see Fig. 96): Taking the upper floor first, the external force P, may be considered as producing equal horizontal reactions at the bottoms of the portal legs, or at the floor level, equal to $\frac{P_1}{2}$ each. A wind moment M is also produced at this floor level, or,

$$M = P_1 h_1, \text{ where } h_1 = \frac{H_1}{2}.$$

Owing to the rigidity of the framework, this wind moment will be resisted by the resisting moment of the column sections, and by the portal connections at the floor-line. This resisting moment must equal $\frac{fI}{y_1}$, where $f =$ the unit-strain on extreme fibres, $y_1 =$ distance of extreme fibres from the neutral axis, and I = moment of inertia of the section. But $M = P_1 h_1$ hence $f = \dfrac{P_1 h_1 y_1}{I}$.

I will be slightly different on the two sides of the neutral axis. On the compression side of the bay, I will be taken as

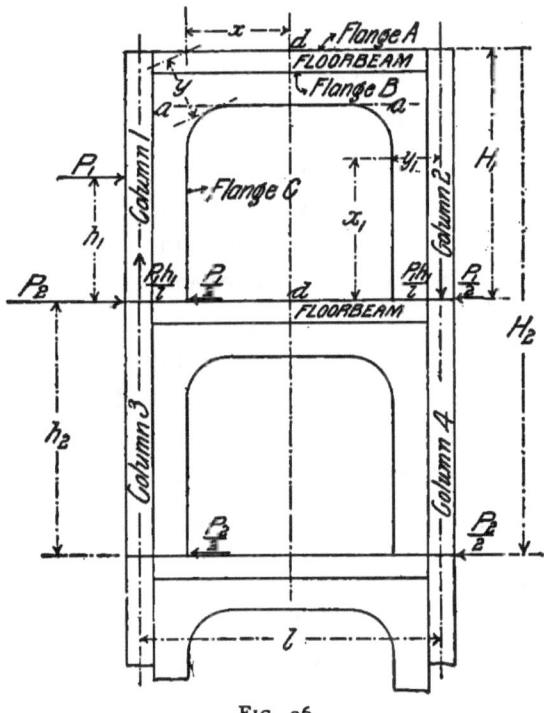

FIG. 96.

the moment of inertia of the section of the column and the portal, while on the tension side, I must be taken for a section of the column and the bolts securing the portal to the floor-beam or to the portal below. If a splice occurs in the column on the tension side, I must be taken for the sections of the bolts connecting the cap-plates of the column, and for the bolts through the portal and floor-beam.

The decrease of load on the one column, and the equal increase in load on the other column will be as before, or

$V_1 = \frac{P_1 h_1}{l}$. In column 2, the vertical column load V_1 due to wind, must be added to the regular column load, the same as in previous discussion. V_1 must also equal the shear on all vertical planes.

The horizontal shear along the line $aa = P_1$, while the horizontal shear in either leg or portal or at bottom of leg $= \frac{P_1}{2}$. These shears will determine the thickness of the webs. The connections of the portal to the column on either side must equal the total vertical shear.

Taking moments about the line dd, it will be found that $\Sigma M = 0$. That is, there is no bending moment along the line dd, and neither the floor-beams nor portals are strained by bending moment along this line.

For a maximum stress in the flange C take a point in flange A, distant x from line dd, and distant y, at right angles, from flange C. Then x times the vertical shear divided by $y =$ stress at section taken, and this is maximum when $\frac{x}{y}$ has its maximum value. The stress in the flange A may be obtained in a similar manner.

The leg of the portal, including column 2, may also be taken as a cantilever, with the two forces $\frac{P_1}{2}$ and V_1 acting on it. The flange C will be in compression, the column itself acting as a tension chord. Assume a point on the centre line of the column, distant x_1 from bottom of leg, and at distance y_1 from the flange C, at right angles. Then $\frac{P_1}{2} \frac{x_1}{y_1} =$ strain in flange C, and this is maximum when $\frac{x_1}{y_1}$ is maximum. There is a slight error in this treatment, but it is on the side of safety. If the flange C is proportioned for these maximum stresses, the requirements will be fulfilled.

In the second story from the top, $V_2 = \dfrac{P_2 h_2}{l}$, considering $P_2 = 2P_1$, or that the stories are of equal height. The concentric load V_1 in column 2 from the column above, and its equal reaction, may be omitted in a calculation of the

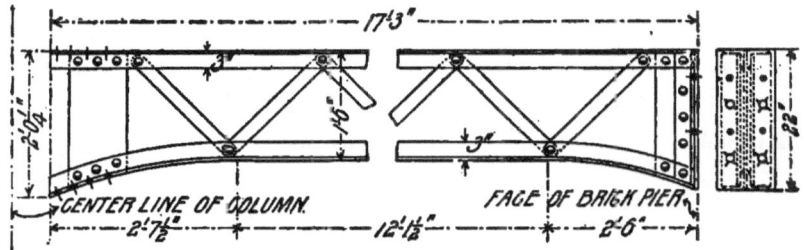

FIG. 97.—Portal-strut used in the Monadnock Building.

strength of the portal bracing (as they are applied along the same straight line), as may also the equal negative effects in column 1.

The vertical shear in this second-story bracing will

FIG. 98.

equal $S_2 = V_2 - V_1$. The horizontal shear across the top of the portal $= P_2$, while in either leg the shear $= \dfrac{P_2}{2}$.

One of the first attempts at a portal system in building

construction was through the use of a portal-strut used in the older portion of the Monadnock Building, as in Fig. 97.

The portal system (3) was used in the Old Colony Building, Chicago, completed in 1894. The portals are placed at two planes in the building—a cross-section of one set being shown in Fig. 98. Wind pressure was figured at 27 lbs. per square foot on one side of the building at a time. Each portal was calculated independently for the sections of both top and bottom flanges, thickness of web, cross-shear on rivets connecting the curved flanges, and for

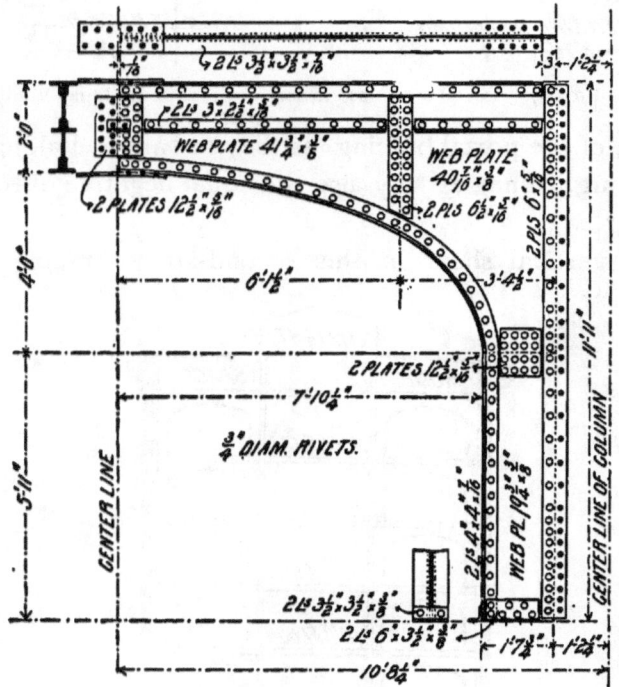

Fig. 99.—Detail of Portal, Old Colony Building.

all splices and connections. A detail of one portal is shown in Fig. 99. This arrangement of wind bracing proved very satisfactory in all respects, and, according to the designer, was cheaper in the end than the sway-rods provided in the first design; but the writer would question whether portal bracing can be provided cheaper than tension-rods, as claimed. With good details in connections and proper regard for their location in the original planning of the building, sway-rods can be used without great expense or trouble. The portal arrangement certainly makes a fine interior appearance if the arched openings are given a slight decorative treatment in plaster, as was done in the Old Colony Building. The floor plan will generally govern the use of either one or the other system, whether the rooms are to be connected by large openings or small doorways.

KNEE-BRACES (4).

The system of knee-braces, or arrangement (4) for wind bracing, is not an economical method, as it produces

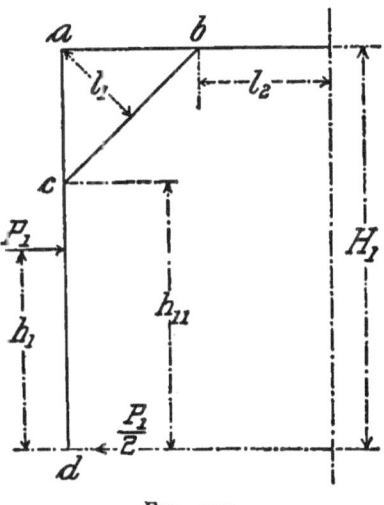

FIG. 100.

heavy bending moments in both the horizontal struts and in the columns themselves. This system may be analyzed as follows (see Fig. 100)

The shear at the top-floor level will be $\frac{P_1}{2}$ at each column.
Then as before, $V_1 = \frac{P_1 h_1}{l}$.

The tension in the brace cb is nearly

$$T_1 = \frac{P_1}{2} \cdot H_1 \cdot \frac{1}{l_1} = \frac{P_1 H_1}{2l_1}.$$

There will be an equal amount of *compression* in the opposite brace. This suggests the use of knee-braces capable of resisting both compression and tension. There will be a bending moment at C whose value is approximately $M = \frac{P_1}{2} \cdot \frac{h_{11}}{2} = \frac{P_1 h_{11}}{4}$. The factor $\frac{h_{11}}{2}$ is used, as the column is considered as square-ended and fixed by the static load

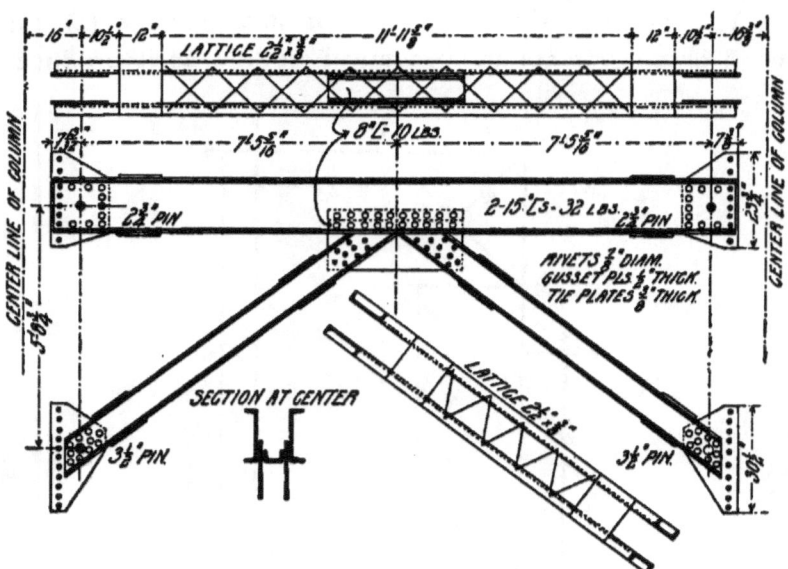

FIG. 101.—Knee-bracing used in the Isabella Building.

and by bolts. This bending moment will also exist at d.

At b there will be a bending moment $M_1 = V_1 l_2 = \frac{P_1 h_1 l_2}{l}$.

This type of wind bracing was used in the Isabella

Building, by W. L. B. Jenney, architect, as shown in Fig. 101.

A modification of the knee-brace system of wind bracing was employed in the new Fort Dearborn Building (1894-95,) by Jenney & Mundie, architects, Chicago. In this case a wind load of 40 lbs. per sq. ft. was taken, and the assumption made that 25 per cent of this wind load would be resisted by the rigid connections provided between the columns and the floor system, leaving 75 per cent, or 30 lbs. per. sq. ft., to be taken up by the exterior columns. This was done by using channel girders between the columns in

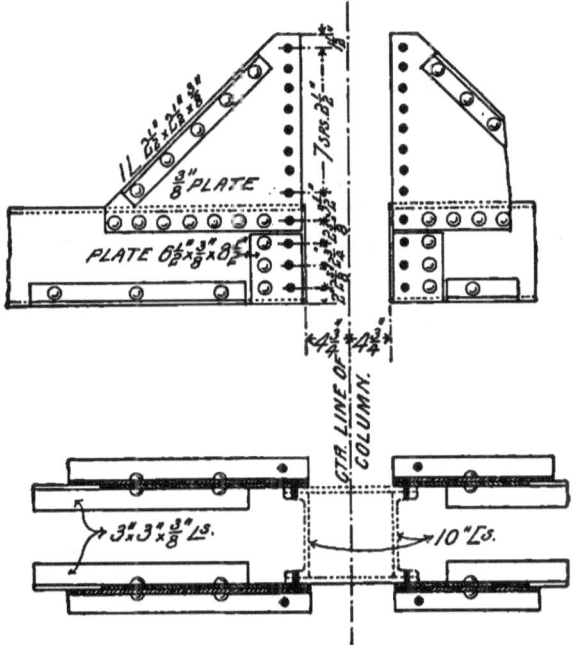

FIG. 102.—Gusset-plate Bracing used in Fort Dearborn Building.

the exterior walls, with gusset-plate connections to the columns, as shown in Fig. 102, 10 in. and 12 in. channels being used generally. In the lower stories, where the wind moment necessitated it, a double system of gusset connections was used, under and above the channel girders,

A somewhat similar method was used in a building in New York City, 120 ft. in height and 24 ft. frontage, designed by L. de C. Berg. The Z columns were used, spaced 12-ft. centres, and anchored to foundations. At three levels in the building occur riveted girders in the exterior walls; the girders connect to the columns by large gusset-plates. At these levels, diagonal ties of flats are also run horizontally over the floor system. An additional load of 15 lbs. vertically was figured in the columns and girders for the effect of the wind. A similar system of horizontal flats was also used in the old Monadnock Building, Chicago. In the Reliance Building the wind strains were transferred from story to story on the table-leg principle. 24-in. plate girders were used in the exterior walls at each floor level, as in Fig. 73.

The effects of earthquakes would scarcely seem to warrant much consideration in our latitude, though Quimby and those who discuss his article, give considerable prominence to it. "The only safeguard against an earthquake is a system of bracing with some elastic material of positive strength, that will so unify a structure that it will hold together, even to the point of overturning bodily."

The Chicago skeleton construction has been adopted in San Francisco, where the fear of earthquakes has, heretofore, been sufficient to keep investors from erecting high buildings. The new Chronicle building and the Croker and Mills buildings are of the Chicago type, twelve stories and over in height, and have served as precedents in that locality.

From a consideration of the wind strains in a building it would seem that a seventh point should be added to the list of headings under the discussion of columns, namely:

7. *Column Joints.* — "The stability of the individual columns in a framed structure is an element of resistance of

considerable value if the connections are rigid," and " wherever adequate rod-bracing is not provided, join the columns by *complete splices*, making them continuous, each column a unit, to fail only by breaking or bending."

Although Quimby seems to limit the necessity of such continuity to cases in which no wind bracing is provided, the writer believes that the method of column joints at each and every floor level is wrong, whether wind bracing be provided or not, and that the tendency should rather be toward design with continuous columns, and riveted members for the main girders and spandrel sections in the walls. Nor should efficient wind bracing be neglected even with these additional factors.

Columns have generally been of single floor lengths, with $\frac{3}{4}''$ cap-plates on top, with the beams connected to the columns by rivets through both top and bottom flanges, those through the bottom flange passing also through the bed-plate and the angle riveted to the column beneath (see Fig. 78). Connections to the bed-plate only should always be avoided, as the lateral strain to be resisted should go to the column and not to the bed-plate. The columns are usually connected to each other by at least four rivets, spaced on opposite sides, as far from the centre of the column as possible, and passing through the cap-plate and connection-angles of each column. If this is done, every rivet driven tends to stiffen the connection of the columns. If the girder loads are heavy, bracket angles must be provided in the lower column to take the shear off the cap-plate. At least $3\frac{1}{2}$-in. bearing in full is given to each beam, and the columns should be carefully planed on the ends, and at true right angles to the column axes.

This method of bracketing the tiers of columns together by means of angles or bent plates, gives a detail that is sufficient to prevent lateral displacement, but because of the elasticity of the brackets in bending, and the large ratio of the height of the column to the base, contributes very little to the rigidity of the structure. The overturning or lift on the windward side is almost always less than the resistance due to dead weight; but the shear is liable to be overlooked, tending, as it does, to topple over all of the columns of a story. The column connection described is not stiff enough to prevent a slight movement, which can be prevented by wind bracing only; and even with wind bracing, it introduces a weakness of the column at the floor level, which the writer believes can be obviated by continuous columns.

In the Masonic Temple, the use of columns of two-storied lengths was tried, as an additional factor of stiffness in so high a building, with the joints "staggered,"

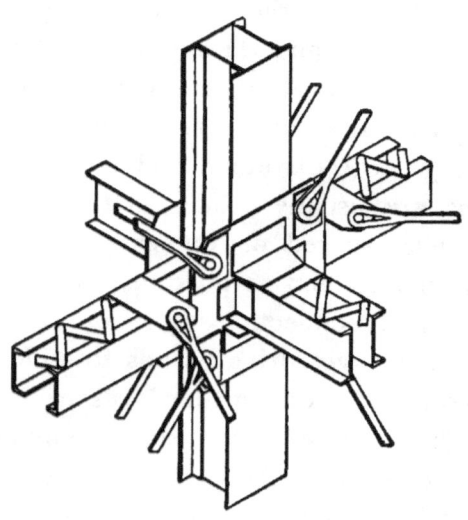

FIG. 103.

or each column breaking joints with its neighbor. The next step was to discard the bed-plates entirely, using *vertical* connection-plates for all column splices. Fig. 103

shows a column splice with connections for the floor-girders and wind bracing, used in the new Pabst Building, Milwaukee, by S. S. Beman, architect. The floor-girders are made of latticed channels, and the sway-rods are connected to the vertical splice-plates of the columns much as the laterals in bridge-work are connected to the chords.

The following clauses relating to the splicing of the Gray columns used in the Reliance Building are from the specifications for the steelwork: "The columns will be made in two-story lengths, alternate columns being jointed at each story.

"The column splice will come above the floor, as shown in the drawings. No cap-plates will be used. The ends of the columns will be faced at right angles to the longitudinal axis of the column, and the greatest care must be used in making this work exact. The columns will be connected, one to the other, by vertical splice-plates, sizes of which, with number of rivets, are shown on the drawings. The holes for these splice-plates in the bottom of the column shall be punched $\frac{1}{8}$ in. small. After the splice-plates are riveted to the top of the column, the top column shall be put in place and the holes reamed, using the splice-plates as templates.

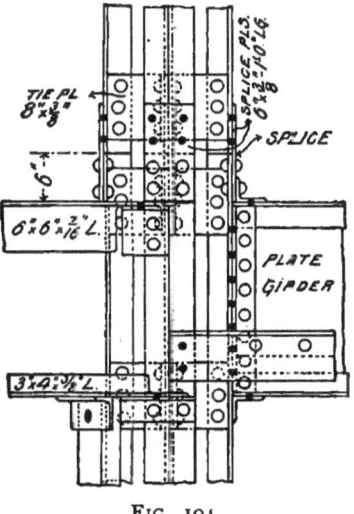

FIG. 104.

The connections of joists or girders to columns will be standard wherever such joists or girders are at right angles to connecting faces of columns. Where connections are oblique, special or typical details will be shown on the drawings."

Fig. 104 shows a typical detail of a column splice in the Reliance Building, where the framing for a bay window joins the column.

Foster Milliken, in his discussion of Quimby's article, classifies the points constituting a perfect joint as follows:

1. Continuity of column from cellar to roof.
2. Proper connections for load and proper distribution.
3. Facility of connections for wind bracing.
4. Ready alignment.
5. Simplicity of design, facilitating erection.

He adds that the ideal column would be one tapering uniformly, with the section varying from floor to floor with the loads, advocating the continuous Phœnix column with pintle-plate connections, as before described. Any such system as this, demanding *built sections* of plates and angles for girders, instead of the conventional rolled beams, would certainly give much more efficient connections with the columns; and joints may be designed adding greatly to the rigidity of the structure, even where the regular transverse bracing is omitted, or where it interferes seriously with the necessary openings in the partitions.

For the theoretical limit in the height of a building, considering the wind pressure, we may assume that the wind acts against the building in a horizontal direction, so that the structure may be taken as being under the same conditions as a uniformly loaded beam, fixed at one end and with the other end free. If this were actually the case with a steel beam, we should make the depth of the beam such that it would deflect less than the amount necessary to crack the plaster. If the beam were supported at both ends, this depth would be one twentieth of the span.

The lengths under these two conditions, to secure the same deflections, must bear the relation one to the other as 0.57 to 1.

If, then, we have an office building or any skeleton structure 25 ft. wide, and make the height twenty times the width, the building would be 500 ft. high, and reducing this in the above ratio, we have 285 ft.

This height would give a theoretical deflection of some 8 in. or 9 in., which would throw the centre of gravity of the upper wall beyond the outer edge. The maximum allowable deflection would be about $2\frac{1}{2}$ in. or 3 in., and this would give a height of from 70 to 95 feet.

The load effect on a uniformly loaded cantilever is four times that for a uniformly loaded beam supported at both ends. If we work on the assumption that the building is analogous to the cantilever beam, and make its height one fourth as great as we would if it were supported at both ends, we should have the depth to the length about as 1 to 5. This would give a height of 125 feet.

Some recent experiments, however (see *Engineering News*, March 3, 1894), on the deflections of tall skeleton construction buildings in Chicago, tend to show that any actual deflections in well-designed and carefully constructed buildings, under very heavy winds, are far less than any theoretical assumptions. Two sets of tests were made, one on the Monadnock Building of seventeen stories, and the other on the Pontiac Building of fourteen stories. Observations were made with transits set in sheltered positions, and these observations were checked by means of plumb-bobs, suspended in the stair-wells from the top floor.

The vibrations in the Monadnock Building from west to east, or in its narrow direction, were from $\frac{1}{4}$ in. to $\frac{1}{2}$ in. The plumb-bob test, however, showed the greatest variation to be in a north and south direction, or longitudinally; but as the walls in three of the four separate divisions of this building are of solid brickwork, from 3 ft. to 6 ft. in thickness, and the length is several times the breadth, it is difficult to believe that any actual longitudinal deflection could be detected.

In the transverse deflections the transits showed a greater deflection in the veneer portion of the building than in the more solid parts, as would very naturally be expected. The time of a complete vibration was two seconds.

The experiments on the Pontiac Building, which is of the veneer type, compared very closely with those on the Monadnock Building, except that the amplitude of the vibration was less in the former building, due to its somewhat more sheltered position. The same peculiarity of an apparently greater longitudinal vibration was noticed here also. The wind was from the northwest, and registered eighty miles per hour.

CHAPTER IX.

PARTITIONS—ROOFS—MISCELLANEOUS.

PARTITIONS.

MOST of the partitions now placed in Chicago office buildings are made from the same character of hollow tile as is used in the floors and around the columns, except that a soft tile is almost invariably used to allow the driving of nails in placing the door-frames and transom-lights. Tile blocks are used in this construction, varying in thickness from 2 in. to 6 in., but the 4-in. blocks are generally used. They may be either square or brick-shaped, and are frequently clamped together, but are always laid to break joint. At all openings in the partitions wood frames are set to stiffen the jambs, and to afford grounds for the plastering, as well as to serve for the attachment of the architraves. The plastering is applied directly to the tile surface.

These partitions may be readily torn down and shifted to suit the tenant, without injuring the construction of the floors or walls. They are never used to sustain loads.

An effective partition which has been used quite extensively, consists of metallic lathing wired to light channel-irons, spaced as studs. Each side is then plastered, making a partition only 2 in. thick. This type of partitions was adopted in the Armour school and in the Dexter office building in Chicago.

Another method is to use $1\frac{1}{4}$-in. I beams spaced as studs 2 ft. on centres. The spaces between these supports are filled in with scratch-coat mortar, and a coat of plastering may then be given each side. If either of these systems of metal studs is used, a strong solution of alum-water should be given the rough coat of plastering to prevent the staining of the finished plaster.

ROOF CONSTRUCTION.

The roof construction in such classes of buildings as are here being considered, should be as thoroughly fire-proof as any other part of the structure. This is secured through the use of tile arches, as in the floors, or by means of book-tile supported on T irons, placed about 18 in. centres. The T irons are supported on I-beam purlins, and if this type is employed care must be taken to see that such a form of book-tile is used as will effectually protect the under surfaces of the T irons. A common method has been to place the book-tile on the flanges of the T irons, thus leaving the lower surface of the T's with a coating of plaster only. Book-tile are now made which project below the metal work, as do the floor arches, thus offering a coating of clay as protection against heat (see Fig. 105).

FIG. 105.

Tile arches of the segmental pattern are often used in roof construction, and the whole is then covered with a layer of concrete which receives the composition roofing (see Fig. 57). The supporting girders and purlins should also be covered, either with special forms of tile blocks or slabs, or else with expanded metal lath to receive a thick coat of cement plaster.

Great care should be taken to see that all spaces between

roofs and suspended ceilings are rendered fire-proof in all their parts, that the spread of unseen fire may be made impossible. Much may be done through a judicious use of metallic lath secured to a light iron framework, in the innumerable instances where a masonry or tile protection becomes impossible.

SUSPENDED CEILINGS.

Such ceilings are usually made of book-tile or of a thin fire-clay tile supported by light T irons. Ceiling tile is often made not over $\frac{1}{2}$ in. in thickness, with grooved edges that fit into 1 × 1 inch T irons, spaced 12 inch centres, which are supported in turn by 3 in. T's hung from the roof purlins.

FURRING TILE,

to take the place of the wood and lath furring used in ordinary construction, is employed to prevent the penetration of the moisture through the exterior walls. These tiles are made similar to the partition tile, and should always be provided with an air-space, to insure a circulation of air, that the injurious effects of damp walls upon the interior finish may be overcome.

FIRE-PROOF VAULTS.

The old system of building brick vaults in tiers is not followed in the modern office building. The vaults are now built of tile and placed as may be desired according to each floor plan, much as the tile partitions. They are not usually shifted, but should it be required, the operation would in no way affect the floor or load-bearing construction. The tile walls should be of considerable thickness, with at least two air-spaces, and the top should also be made of two thicknesses of tile in case the vault does not run to the ceiling.

STAIRWAYS AND ELEVATOR ENCLOSURES.

The stairways are usually made of cast risers, strings, and newel-posts, with wrought railings and wooden or polished bronze or brass hand-rail. All exposed parts of the risers and strings are generally specified to be panelled

FIG. 106.—Main Entrance and Elevator-hall, Marquette Building.

and ornamented as per detail drawings, and provided with lugs and flanges to receive the marble treads and platforms. The metal-work for the stairways and the elevator guards or enclosures are heavily electroplated in brass, copper, or bronze. An aluminium finish was tried in the newer portion of the Monadnock Building, Chicago, but it has tarnished very badly. Fig. 106 shows the main entrance-hall to the Marquette Building, serving as a good illustration of the decorative treatment which may be given the columns and exposed or sunken girders in the ceiling. Fig. 107

FIG. 107.—Entrance-hall, New York Life Insurance Building.

shows the main entrance to the New York Life Insurance Building, with the walls, ceiling, and stairs finished in Italian marble and the floor of mosaic.

The main stairway and entrance-hall to the Fort Dearborn Building are given in Fig. 108.

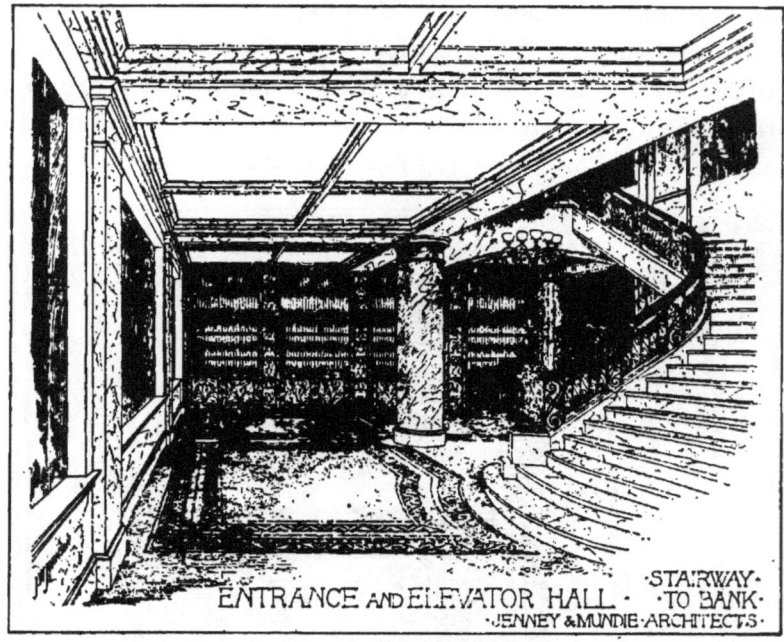

Fig 108.—Entrance-hall, Fort Dearborn Building.

COLUMN-SHEETS.

Before the column-sheets may be started it is necessary that all loads occurring in the structure be definitely settled. These loads include, as suggested in the previous chapters, the weights of all structural material (floors, roof, piers, spandrels, and the like), besides wind, snow, elevator, and tank loads. The column-sheets may then be started, forming a tabulated list of all the loads transferred to the footings through the columns. From these sheets may be seen the approximate load that each column must carry at any floor, starting with the upper-story columns, supporting the roof load only, and adding in the loads at the successive floors down to the foundations. The column weight itself is first assumed, and then corrected, after the proper section is obtained.

The column-sheet used in the Masonic Temple calculations was as follows:

		Column 1.		Column 2.	
		Load on Column.	Load on Footing.	Load on Column.	Load on Footing.
Roof.	Floor load............ Masonry piers......... Spandrels............. Elevators............. Tank loads........... Weight of column..... Total.............				
20th Floor.					

The column-sheet used in the Venetian Building was made as in the accompanying table:

Column 1.	Roof.	Attic.	12th Floor.	Basement.	Total.
Load from column above........ Floor load, dead............... Floor load, live............... Spandrels..................... Elevator loads................ Estimated weight of column..... Total.......					
Wind Loads.					
Concentric wind loads........... Eccentric wind loads........... Total wind load...........					
Column 2.					
Etc.					

The following column-sheet is to be recommended as combining all requisites in a tabulated statement:

		Column 1.			Column 2.
		Load on Column. Concentric.	Load on Column. Eccentric.	Load on Footing.	
ROOF.	Floor load............ Masonry piers......... Spandrels............ Elevator loads........ Tank loads, etc........ Weight of column...... Wind................. Total........				
	Area required for col...	sq. in.	sq. in.	Foot'g area	
	Material of column.....			sq. ft.	
		Load on Column. Concentric.	Load on Column. Eccentric.	Load on Footing.	
16TH FLOOR.	Floor load............ Etc.				

The final loads on the basement columns taken from these sheets will show the loads for which the footings themselves must be figured, while the final loads on the footings will give the weights for which the clay areas must be proportioned.

CHAPTER X.

FOUNDATIONS.

No part of the work of the engineer requires more care and skill than the design and execution of the foundations. " Where it is necessary, as so frequently it is at the present day, to erect gigantic edifices—as high buildings or long-span bridges—on weak and treacherous soils, the highest constructive skill is required to supplement the weakness of the natural foundation by such artificial means as will enable it to sustain such massive and costly burdens with safety " (Baker). And nowhere is this more true than in Chicago, where it is almost impossible to penetrate to bed-rock with any degree of practicability, and where the soil underlying the city consists of blue clay (below a soft loam or quicksand) at about 12 or 14 feet below the sidewalk grade, and thence down to a bed-rock of limestone from 40 to 80 feet below the street level. The clay is hard and firm in the upper strata, but becomes soft and yielding as it descends, often containing pockets of spongy material, thus necessitating borings for reliable information of particular localities. Borings have been extensively made, both by private parties and by the government, resulting in an allowance of from $1\frac{1}{2}$ to 2 tons per sq. ft. on the clay, with due consideration for proper settlement. Baker states as follows on this subject : " The stiffer varieties of what is ordinarily called clay, when kept dry, will safely bear from 4 to 6 tons per sq. ft., but the same clay, if allowed to become saturated with water, cannot be trusted to bear more than 2 tons per sq. ft. At Chicago the load ordinarily put on a thin layer of clay (hard above and soft below, resting on a thick stratum of quicksand) is 1½ to 2 tons per sq. ft., and the settlement, which usually reaches a maximum in a year, is about 1 in. per ton of load."

Unequal settlement is thus the great evil that must be guarded against, for settlement will come, slowly but surely, and in all good designs it is provided for in the start by making the structure some 3 in. to 5 in. higher than its final level. The evil of unequal settlement can hardly be better exemplified than in the case of the United States Government Post Office and Custom House in Chicago, built in 1877, and now about to be replaced by a new one. The foundations consist of a continuous sheet of concrete, made in different layers, but altogether 3 ft. 6 in. thick. Some portions of the building were extraordinarily heavy, others comparatively light, but the Washington architects thought the concrete sufficient, even though there were bad sloughs under the building. But it has proved a most dismal failure, and even a menace to life and limb. It has settled nearly 24 in. in places, and a dropping of some part of the structure is no unusual occurrence. After but eighteen years of service this example of government architecture and engineering has been known as "The Ruin" in Chicago and vicinity.

The investigation of the compressibility of the soil leads to the conclusion that, if we wish to procure uniform settlement, all parts of the foundation areas must be exactly proportioned to the loads they have to carry. Examples are not lacking, in Chicago and elsewhere, of the actual crushing of light piers, when alternating with heavy ones, because, *proportionately*, the lighter piers had too great a footing area. In the Mills Building in New York City the mullions in the lower floors of the building and over the light foundations were seriously damaged and even crushed, because they were not strong enough to force down the lighter piers of too large an area, as fast as the heavy piers were settling.

The footings themselves must be of sufficient strength to distribute the applied loads over the requisite area; in this way only can satisfactory results be obtained.

The arrangement of independent piers was first advocated in Chicago by Frederick Bauman, in a pamphlet published by him in 1872, entitled "The Method of Constructing Foundations on Isolated Piers," and this method has certainly been brought to a high degree of perfection by the engineers of Chicago. The rapid development of foundations is well exemplified by the great change in methods employed at the site of the Woman's Temple. In 1890 the lot where this building now stands was bought by the present owners. Extensive masonry foundations had previously been built here for a structure that was never erected, and upon the preparation of plans for the Temple, the first thing done was to remove these massive masonry piers at a cost of many thousands of dollars. The old system consisted of stone piers made of successive layers of large stones, stepping out until a sufficient base was obtained. One of these newer "raft" footings is here given, and also one of the old masonry type (Figs. 109, 110).

The objections to these old piers were many: they were bulky, occupying too much space; they were heavy and costly as regarded the time necessary for building; and the allowable offsets of the masonry-work seriously limited the load-bearing surface of the clay.

These piers in the Woman's Temple were all underlaid with a bed of concrete, resting on the clay stratum about 10 ft. below street grade. A comparison of some of the above points may be made as follows:

I. *Space.*
   1st. Top of concrete to bottom of casting = 1′ 8″.
   2d.   "   "   "   "   "   "   "   = 7′ 0″.

Or, comparing the parts above the common bed of concrete,
   1st = 217 cu. ft.,   2d = 691 cu. ft.

This point of space is a very important one, as has been before mentioned, since basement-space is now quite as valu-

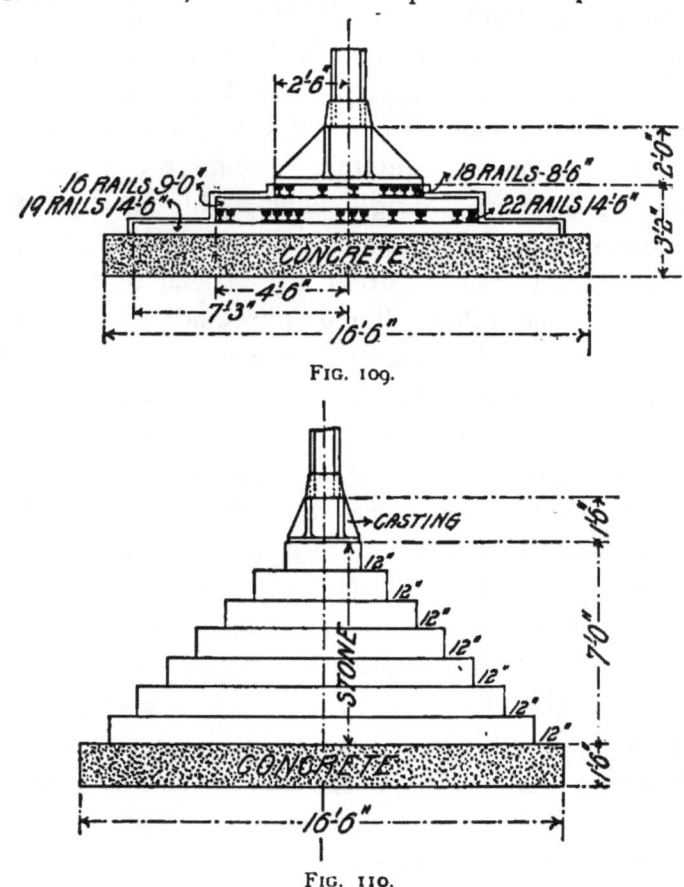

Fig. 109.

Fig. 110.

able as any office-space, for use as restaurants, cafés, or for the large boiler and electric-light plants necessary. Indeed, it is of frequent occurrence to extend the basement-space out under the sidewalks and even alleys. Thus, to gain cellar-room, the foundation must either be lowered or made thinner. The first has been ruled out of Chicago practice, because it has been clearly demonstrated that the less the clay stratum is broken, the more uniform and satisfactory the settlement will be. Hence the present details.

II. *Weight.*—Rating the masonry at 150 lbs. per cu. ft., concrete at 140 lbs., and allowing 44 cu. ft. for the steel in No. 1, the weights are:

$$\text{No. 1} = 103,000 \text{ lbs.}$$
$$\text{No. 2} = 261,000 \text{ lbs.}$$

The load for this foundation is 800,000 lbs. and the saving in weight, through the use of the raft foundation, is thus sufficient to allow an additional story, without adding to the load on the clay. On large foundations this difference is still greater. In the case here assumed the 103,000 lbs. is about 13 per cent of the load carried, but in some cases, under very heavy loads, it has been found to run as high as 20 per cent (see "Steel Rail Foundations," *Engineering News*, August, 1891).

This saving in weight is one of the factors that makes our highest buildings possible, and even the "skyscrapers" are not loading the clay as severely as some of the older structures. When the foundations of the old masonry building were torn out to make room for the new Reliance office building, the clay was found to be loaded to 2 tons and over per sq. ft. for a five-story building, while "The Fair" Building is loaded to 2850 lbs. only, for a sixteen-story modern structure.

III. *Cost.*—In general, the cost of stone foundations will be less than iron ones, but considering the renting-space in basements, this difference will be quickly made up where the latter are used.

IV. *Time.*—In the time required for building operations the new foundations are greatly superior, as rails and beams are easily obtained and cheaply handled.

V. *Load-bearing Area.*—As to the fifth point, stone foundations under side walls frequently cannot step out sufficiently to get the proper bearing area without projecting into the next lot. But with iron we can combine several footings, or use cantilever foundations, thus securing the desired results.

As may be seen by Fig. 109, the new type of foundations consists first of a layer of concrete, about 2 ft. thick, upon which come layers of I beams or rails, each layer laid transversely to those just below or above. The spaces between the rails are rammed tight with concrete, which preserves the iron from the action of air and water.

It is the judgment of the best engineers that the area of the foundations on the clay should be proportioned to the dead loads only, and not to theoretical or occasional loads. Whenever live loads have been figured on both the interior columns and on the columns in the exterior walls, the exterior columns have always been found to settle more, from the fact that the live load forms a larger percentage of the interior-column loads than of the wall-column loads. Experience has also shown that after the clay has been compressed by a load of 3000 lbs. per sq. ft., and allowed several months' repose, no very perceptible addition to that compression will result without a material addition to the load. So that it is good practice to neglect live loads on the clay for hotels, office buildings, or lightly loaded retail stores. In warehouses, however, or in buildings carrying very heavy permanent or shifting floor loads, or machinery in motion, the change of loads and the jarring increase the compression of the clay very largely. Hence we must make extra allowances in such instances.

In all cases where live loads have been figured on the columns, consistency requires that whatever loads have been figured on basement columns, must be figured on the metal in the foundations; or the clay areas are proportioned for dead loads only, while the strengths of the foundations themselves are figured for dead + some live load. But, as before said, many of the best buildings have entirely disregarded live loads on the footings. W. L. B. Jenney advocates as follows: In hotels, office buildings, and retail stores neglect the live loads on the footings, but figure them in heavy warehouses, machinery plants, etc. Where much pounding occurs, as in machinery in motion, use double the weight as dead load that we figure for live load.

In "The Fair" Building, where a large quantity of merchandise is stored, and the aisles are constantly filled by throngs of people, the following system was used: The floor-beams carry all the dead + live loads, the girders carry the dead load + 90 per cent of the live load, while any one column carries a percentage of the sum of the live loads of all the stories above that column + the total dead load. The percentage of live load is given in the last column of the accompanying table:

Column.	Live load on beams.	Per cent for column.
Attic	——	100 per cent.
16th story	75 lbs.	90 " "
15th "	75 "	$87\frac{1}{2}$ " "
14th "	75 "	$77\frac{1}{2}$ " "
13th "	75 "	$72\frac{1}{2}$ " "
......	......	Decrease of $2\frac{1}{2}$ per
......	......	cent in each story.
6th story	75 lbs.	55 per cent.
5th "	130 "	$52\frac{1}{2}$ " "
Basement	130 "	40 " "

No live load was figured on the clay area, but the allowable pressure per square foot was taken at a very conservative figure—2850 lbs.

The first use in Chicago of iron rails, in connection with masonry or concrete footings, occurred in the Montauk Block, by Burnham & Root, architects. The old method of pyramidal foundations of dimension stones was used with a concrete base 18 in. thick. Iron rails were built into this concrete to obtain a larger offset than could otherwise have been obtained.

At first old rails were employed in these foundations, but now practice demands as reliable material in this portion of the metal-work as in any other. Steel rails at 75 lbs. per yd. are generally used unless steel beams are required. Ordinarily, rails are cheaper than beams; more iron is required, but at less cost than in beams. The concrete, too, is easier to ram between the rails, and the webs are always thick.

Under very heavy loads, or long spans, beams become necessary, 10 in. to 20 in. I beams being frequently used. Only the projecting portions are strained as beams, hence the place for beams is at the top of the pile—the larger the proportion of iron or steel uncovered the more economical the foundation. 20,000 lbs. and 16,000 lbs. have been used for fibre strains in steel beams and iron rails, though the new Chicago ordinance limits the fibre strain to 14,000 lbs. and 11,000 lbs. per square inch. In the Old Colony Building the steel beams in the foundations are strained to a fibre strain of 14,000 lbs. under the dead weight of the building alone, while the maximum dead plus live loads induce a fibre strain of 21,000 lbs. per sq. in. of extreme fibre. The Carnegie strike at the time of building precluded the possibility of obtaining heavier beams than 15-in. 90-lb. I beams, so the strain was allowed under the press of circumstances.

RAIL FOOTINGS.

The raft footings as first employed were made of rails only, the usual method of figuring being as follows: The number of square feet of footing required equals $\frac{\text{load on column}}{\text{pounds per sq. ft. on earth}}$. Multiply the result by 250 (equals approximate weight of footing per square foot), add to the original load, and refigure. The layers are then laid off, the projection of any layer beyond the one immediately above being always 3' 0" or less. The moments on the projecting portions of the layers are then found, and these moments, divided by the allowable bending moment per rail, usually taken at 12,500 foot-lbs., give the number of rails required in the different courses. One extra rail is usually added to each layer as a matter of safety.

The following table gives the properties of the rails from the North Chicago Rolling Mills. The 75-lb. rails are most commonly used.

No.	Weight.	Height.	Base.	u.	I.	R.	$\frac{M}{f = 16,000}$
6504	65 lbs.				15.86	6.86	9,150
7501	75 "	$4\frac{3}{4}''$	$4\frac{3}{4}''$		21.00	8.30	11,070
7503	75 "	$4\frac{3}{4}$	$4\frac{3}{4}$	$2\frac{3}{4}''$	21.66	9.37	12,500
8001	80 "	5	5	$2\frac{3}{4}$	26.36	9.99	13,320
8501	85 "	5	$4\frac{3}{4}$	$2\frac{5}{8}$	27.32	10.41	13,880
8502	85 "	$5\frac{3}{16}$	5	$2\frac{5}{8}$	29.22	11.13	14,840
8503	85 "	5	5	$2\frac{7}{16}$	25.38	10.03	13,370

The table following is taken from the footings of the Great Northern Hotel, giving the loads on columns, areas of footings, and the calculated weights per square foot of the rails and the concrete in the footings. All rails were 75-lb. rails, No. 7503 in the previous table. The bottom courses of all footings were of concrete, 12" thick, extending 6" beyond the lower course of rails, but the weights of these concrete courses are not included in the following. Cast shoes 4' 0" × 4' 0" were used under all the columns. The concrete was figured as weighing 125 lbs. per cubic foot.

Load on Col.	Area of Footing, sq. ft.	Weight per sq. ft.	
		Rails.	Concrete.
415,470	12' × 11¾ = 141'	49	83
433,440	10 × 14½ = 146	58	60
435,820	9 × 16¼ = 146	77	83
461,100	12 × 13 = 156	42	80
496,240	10 × 16½ = 163	79	91
526,850	12¾ × 14 = 178	66	82
531,740	13 × 14¼ = 185	60	78
571,360	13¼ × 14¼ = 192	67	74
595,920	12½ × 16 = 200	67	88
621,560	13 × 16 = 208	60	94
637,240	13⅜ × 16 = 214	68	68
666,000	15 × 15 = 225	66	105
672,000	13 × 17½ = 228	67	93

BEAM AND RAIL FOOTINGS.

The next step made in the development of the raft footings was in the use of I beams for the upper course or courses. Fig. 111 shows a foundation which was figured as follows (see *Engineering News*, August, 1891):

The column load was 1,166,000 lbs. The allowable pressure per square foot on the clay was taken at 3,000 lbs., giving a footing 22' 8" × 17' 3". The lower layer of concrete was 18" thick, projecting 8" beyond the lower course of rails. Fifteen-inch steel beams were used in the top course, weighing 50 lbs. per foot. The allowable moment on each beam equalled 117,700 ft.-lbs. The remaining courses were of steel rails, 4¾" high and 4¾" base, 75 lbs. per yard weight, with an allowable moment of 12,100 ft.-lbs. per rail.

In the upper course, as many beams are used as the space under the column casting will allow. The projecting arms must therefore be determined. The total length of the I beams so found will fix the width of the second course from the top, and the projecting arms must be found for this course as in the first case.

The arms of the lower two courses are fixed by the

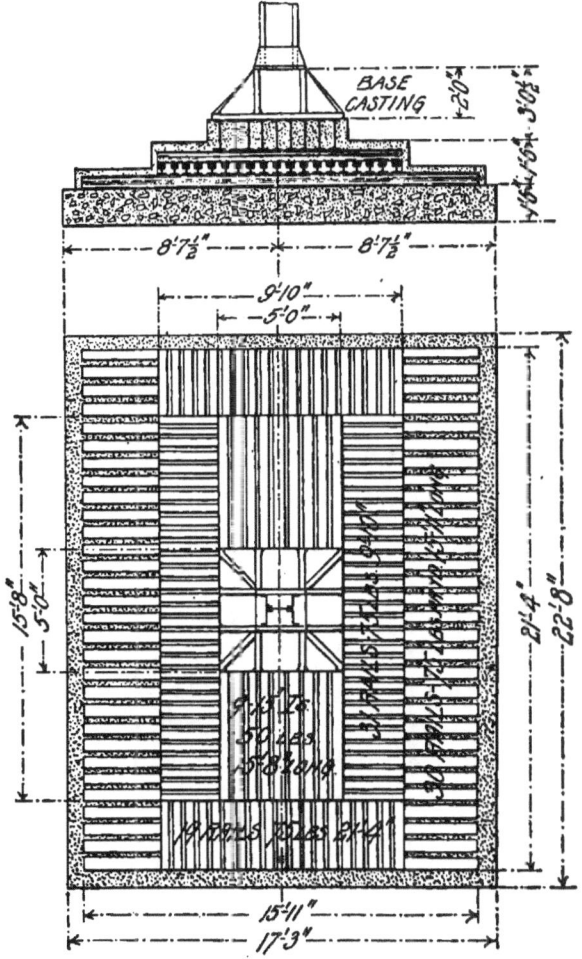

FIG. 111.—Beam and Rail Footing from "The Fair" Building.

lengths of the upper ones, and by the dimensions of the clay area; hence the question is, how *many* pieces are required? The formulæ used may be derived as follows:

Let $y =$ projecting arm in any course;
$a =$ width of supporting area;

l = total load on footing;

M = bending moment on one side of the layer.

Then the length of beam or rail = $a + y + y = a + 2y$. The total load on $y = \dfrac{ly}{a+2y}$, and since the distribution of the load on every layer is uniform, we have

$$M = \frac{ly}{a+2y} \times \text{lever arm } \frac{y}{2} = \frac{ly^2}{2(a+2y)}.$$

In calculating the lower two courses, y becomes a known quantity and M an unknown. In the upper two layers M is given by the number of the beams used and y is unknown.

Considering now the top course, under the base casting, 5' 0" × 5' 0" in area, we find that 9 beams only can be placed under the casting, allowing sufficient space between them for the ramming of the concrete.

M for each beam = 117,700 ft.-lbs. Hence M for the whole layer = 9 × 117,700 ft.-lbs. = 1,059,300 ft.-lbs. Then $\dfrac{1,166,000 y^2}{2(5+2y)} = 1,059,300$, whence $y = 5'\ 4''$. The length of this layer then becomes $5 + 2y = 15'\ 8''$.

For the second course we find that 31 rails spaced about 6" centres may be placed under the 15' 8" beams. Closer spacing than this may be used if necessary. The load now equals 1,166,000 lbs. + the weight of the top course (about 19,000 lbs.). Then $\dfrac{1,185,000 y^2}{2(5+2y)} = 375,100$; whence $y = 2'\ 5''$. The length of the rails therefore = $5'\ 0'' + 4'\ 10'' = 9'\ 10''$.

For the calculation of the lower courses, we know that the area covered by the bottom course is 15' 11" × 21' 4". This leaves a projection of 3' 0½" for the bottom course, and a projection of 2' 10" for the next to the bottom layer.

Then for the third or next to the bottom course, we have

$$\frac{ly \times \dfrac{y}{2}}{a+2y} = \frac{1,200,000 \text{ lbs.} \times 2\frac{5}{6} \text{ ft.} \times 1\frac{5}{12}}{21\frac{1}{3}} = 225,780 \text{ ft.-lbs.} = M.$$

This moment requires 19 rails to be used in the layer.

For the bottom course

$$M = \frac{1,220,000 \text{ lbs.} \times 3\frac{1}{24} \text{ ft.} \times 1\frac{25}{48}}{15\frac{11}{12}} = 343,000 \text{ ft.-lbs.}$$

This requires 29 rails, 30 being used for safety.

It will be noticed in the above calculation that the moments have been taken for the projections of the several courses beyond the adjacent supporting layers only. Thus in the figures for the next to the bottom course, as given above, $y = 2'\ 10''$. If, however, the foundation be taken as a whole, and the bending moment on the third course is taken around the edge of the cast base, the same as the top course was figured, we have $y = 8\frac{1}{6}$ ft., or,

$$M = \frac{1,200,000 \times 8\frac{1}{6} \times 4\frac{1}{12}}{21\frac{1}{3}} = 1,920,000 \text{ ft.-lbs.}$$

This must be resisted by the combined moments of the 9" I beams in the top layer, and the 19 rails in the third layer, or 1,059,300 + 229,900 = 1,289,200 ft.-lbs. This assumption leaves a difference of 630,800 ft.-lbs. which has not been cared for.

The practice in regard to the calculation of such footings is still an unsettled question. Those who used the first method claimed that the action of the concrete filling, with its tendency to bind the iron and concrete together, caused the foundation to act as a whole, and thus possess a moment of resistance much greater than the sum of the resistance of the individual layers. But in view of the uncertainty of any such assumption, the other method of calculating all moments about the edge of the casting would seem more logical, as well as being on the safe side. Both methods are now being used in Chicago buildings.

The use of rails in footings has been succeeded almost entirely by the use of I-beams throughout.

Fig. 112 shows a footing used in the Marquette Building for a column load of 920,250 pounds.

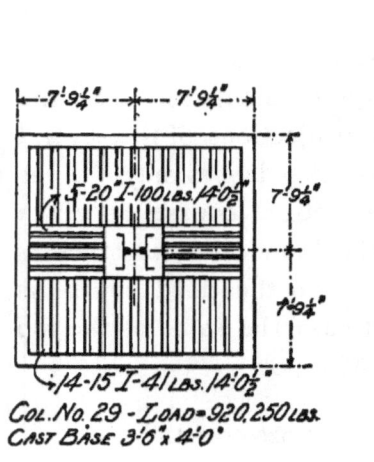

FIG. 112.   FIG. 113.

Fig. 113 is taken from the same building, and is figured for loads of 406,340 lbs. on column 32, and 561,790 lbs. on column 44.

In determining the sizes of the beams or rails in any layer, care must be taken to leave sufficient clearance between the flanges to admit the concrete which must be rammed in place. If stone, broken to pass a $\frac{3}{4}''$ ring be specified, $1''$ as a minimum between the flanges will answer.

In covering these rails with concrete 4 inches of concrete should be left at the ends and sides of the rails, and 1 inch on the tops. A plank frame is made of the same size as the concrete bed, and at the proper height by the aid of levels. After this is filled another frame is made for the next course, and so on. The concrete is made of the best Portland cement, usually 1 part cement to 4 parts of broken stone and 2 parts of coarse sand. The concrete must be well tamped between the beams, and the whole exterior plastered with pure Portland cement mortar, so that no metal-work is exposed. A bed of concrete 18″ or 2′ 0″ thick comes under all, projecting 6″ to 12″ beyond the rails.

COMBINED FOOTINGS.

The raft foundation is particularly valuable where the positions of loads in reference to each other are bad. We may then use *compound* foundations, combining several by means of long beams—as under smokestacks, party-walls, etc.

One of the most delicate problems is the construction of a very heavy building by the side of one already completed, so that the latter will not suffer by settlement, due to the additional weight of the new building.

Such settlement was shown to a remarkable degree in the Studebaker Building, next to the Auditorium, Chicago. The former settled from 10″ to 12″ from the weight of the latter. To obviate such settlement the old wall is carried on timbers, supported at either end by jack-screws. The new wall is then put in, and, with the new foundation which is provided, settles gradually. The jack-screws under the old building are turned as occasion requires, to keep the old wall at its proper level. This is continued until all settlement ceases, when the jack-screws are removed, one by one, and a new wall is substituted under the old building.

If access cannot be had to the basement of the old building, or underpinning, in the manner above described, is impossible, cantilever foundations must be employed. The old foundations must not carry any additional weight, and we cannot substitute new footings; hence the usual type of raft footing is used, but several are combined, and the centre of gravity of the combined area coincides with the centre of gravity of the loads. On these footings come high cast-iron shoes, supporting cantilever girders which carry the columns and wall of the new building, immediately next to the old one, and yet transferring all the load, with the attendant settlement, away from the lot-line. The first cantilever footings introduced in Chicago were used in the Manhattan and Rand-McNally buildings at about the same time. The boilers, etc., in the basements of the adjoining buildings could not be disturbed to allow the introduction of new party-footings, so the cantilever types were adopted for the new structures, and the foundations of the old ones were not disturbed. This method was employed in the Western Union Building in New York, where

Fig. 114.

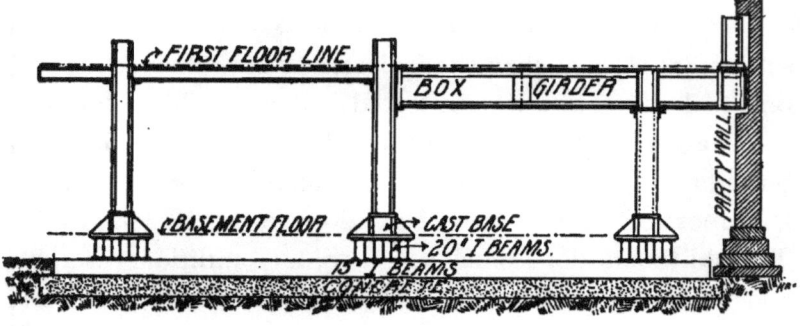

Fig. 115.

a load of 286 tons was transferred from a corner to more secure footings.

Such a combined footing may be analyzed as follows:

Taking Fig. 114 as a plan and Fig. 115 as an elevation, the line of flexure of the 15" I beams will be as in Fig. 116. To find the maximum bending moment on these beams we must

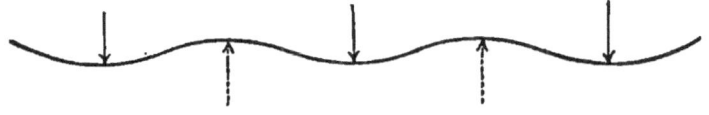

FIG. 116.

compute the various bending moments and compare. The bending moment will be maximum when the shear $= 0$. In this case there are five such sections, as shown by the line of flexure; hence we must compute the moment at each point to find the greatest. The moments under the columns will be $+$, causing convexity downward, while the moments between the columns are $-$, causing convexity upward. Fig. 117 may then be used.

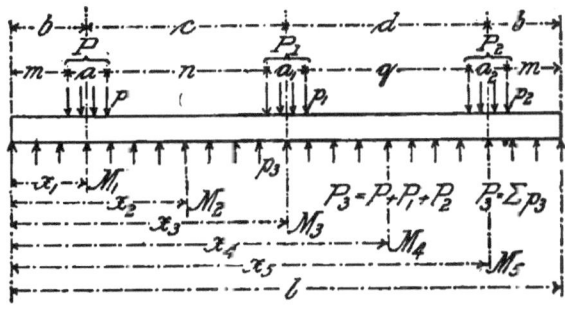

FIG. 117.

To find the distance of the centre of gravity of the loads from the left end we have

$$x = \frac{Pb + P_1(b+c) + P_2(b+c+d)}{P+P_1+P_2}.$$

Let $\dfrac{P}{a} = p$, $\dfrac{P_1}{c} = p_1$, $\dfrac{P_2}{a} = p_2$, and $\dfrac{P+P_1+P_2}{l} = p_3$.

The distances from the left end of the beams to the points

where $S = 0$, or the distances x_1, x_2, x_3, x_4, and x_5, are then found to be as follows:

$$x_1 p_2 = (x_1 - m)p, \text{ or } x_1 = \frac{mp}{p - p_2};$$

$$x_2 p_2 = P, \text{ or } x_2 = \frac{P}{p_2};$$

$$x_3 p_2 = P + [x_3 - (m + a + n)]p_1, \text{ or } x_3 = \frac{(m + a + n)p_1 - P}{p_1 - p_2};$$

$$x_4 p_1 = P + P_1, \text{ or } x_4 = \frac{P + P_1}{p_1};$$

$$x_5 p_2 = P + P_1 + [x_5 - (m + a + n + a_1 + q)]p_2,$$
$$\text{or } x_5 = \frac{(m + a + n + a_1 + q)p_2 - P - P_1}{p_2 - p_2}.$$

The bending moments at these points are readily found by taking the moments of the external forces on one side of the point in question; thus M at the first point (remembering that $M = \frac{Wl}{2}$ for a uniformly loaded cantilever) is

$$M = \frac{p_2 x_1^2}{2} - \frac{p(x_1 - m)^2}{2};$$

$$M_2 = Pb - P\frac{x_2}{2} = P\left(b - \frac{x_2}{2}\right);$$

$$M_3 = \frac{p_2 x_3^2}{2} - P(x_3 - b) - \frac{p_1(x_3 - m - a - n)^2}{2};$$

$$M_4 = Pb + P_1(b + c) - (P + P_1)\frac{x_4}{2};$$

$$M_5 = \frac{p_2 x_5^2}{2} - P(x_5 - b) - P_1(x_5 - b - c)$$
$$- \frac{p_2(x_5 - m - a - n - a_1 - q)^2}{2}.$$

In general cases M_2 and M_4 will be small except where

the columns are very far apart, and the maximum bending moment will be at either M_1, M_2, or M_4, according to which column is the heaviest. If the cast bases are strong enough to carry the superimposed loads on their perimeters, and the long beams form the top course, the values of M_1, M_3, and M_5 will be reduced. M_2 and M_4 would not, however, be altered.

Sufficient deflection could hardly take place to increase materially the reaction under the central column, if figured as a continuous girder; but if so calculated, the clay reaction would be of a varying intensity, as in Fig. 118.

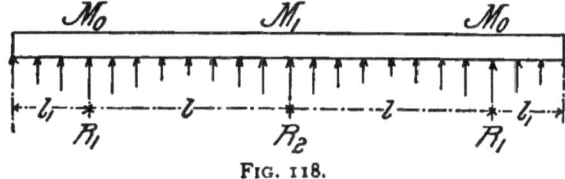

FIG. 118.

Thus, from Clapyron's formula, we have

$$M_0 + 4M_1 + M_0 = \frac{pl^2}{2}$$

for a continuous girder of two equal spans, l. But in the case assumed

$$M_0 = \frac{-pl_1^2}{2} \text{ and } M_1 = -\frac{1}{8}pl^2 + \frac{M_0}{2}, \text{ or } M_1 = -\frac{1}{8}pl^2 - \frac{pl_1^2}{4}.$$

Taking now the shears S_1 and S_2, on the left and right respectively, of the reaction R_1, and remembering that $S_1 + S_2 = R_1$, we have

$$S_2 = \frac{M_1 - M_0}{l} + \frac{pl}{2}, \text{ and } S_1 = pl_1.$$

Then

$$R_1 = \frac{-\frac{1}{8}pl^2 - \frac{pl_1^2}{4} + \frac{pl_1^2}{2}}{l} + \frac{pl}{2} + pl_1$$

$$= \tfrac{1}{8}pl - \frac{pl_1^2}{4l} + \frac{pl_1^2}{2l} + \frac{pl}{2} + pl_1 = \tfrac{3}{8}pl + \tfrac{1}{4}\frac{pl_1^2}{l} + pl_1,$$

where $\tfrac{3}{8}pl$ is the reaction due to the loads on the two spans l, the same as in the regular formula for two spans, and p_1l is the reaction due to the cantilever load, while $\tfrac{1}{4}\frac{pl_1^2}{l}$ is the effect due to the use of the beam as a continuous girder.

Also,

$$R_2 = \tfrac{5}{4}pl - \tfrac{1}{2}\frac{pl_1^2}{l}.$$

These reactions show a varying tendency in the unit pressure on the clay, as in Fig. 118.

In the first example we made the assumption that the reaction from the clay was uniform per foot of length of the footing. According to the law of the continuous girder this would not be true, as we have seen; but when we consider that the beams are generally of sufficient depth to prevent any appreciable deflection, and that the unifying tendencies of the concrete cause the footing to act more or less as a whole, the assumption is undoubtedly justifiable.

SETTLEMENT.

It must not be forgotten that the footings are designed for the final loads that rest on them, and at all stages of the construction the same relation must be maintained between the weights on the various piers that will exist in the completed state, if uniform settlement is desired. This was well exemplified in the case of the Auditorium tower, which extends many stories above the main building, thus bringing greater weights on the tower footings. Here the tower foundations were loaded with varying weights of pig iron at the different stages of construction, in order that the proper relative excess on these piers should be preserved as in the final weight. Even with all these precautions, and after most careful tests of the ground beforehand, this tower has settled more than originally allowed for.

When a test load is applied to the surface, an initial settlement occurs on the surface at a pressure of 1 ton per sq. ft. Another settlement is produced under an increased weight, which ceases in a few hours, and further settlement will not directly occur even with a load of 4500 lbs. to 3 sq. ft. There is, however, a further progressive settlement, owing to the gradual pressing out of the water from the clay. Baker says: "The bearing power of clayey soils can be very much improved by drainage, or preventing the penetration of the water." That the water is pressed from the clay was shown to be the case by careful observations made at the Auditorium. Wells were sunk some 24' 0" deep, 5' 0" in diameter, and 4' 6" from the foundations. The borings were made through the stratified clay, and it was shown that the clay became more and more compact from time to time, thus proving that this squeezing process does take place. The settlements were here carefully watched for a number of years, and they were found to be uniform— about $\frac{1}{18}$" per month.

If the building is heavy, an immediate settlement of from $2\frac{1}{2}$" to 4" is noticed, followed by a gradual progressive settlement. The Monadnock Building 200' high, with 3750 lbs. per sq. ft. on footings, settled 5", while 6" was allowed for. The Western Banknote Building, eighteen stories high, built on quicksand over clay, with solid masonry walls and fire-proofed, settled $2\frac{1}{2}$". The Home Insurance Building settled $\frac{3}{4}$" under two additional stories, and "The Fair" Building settled only 1".

PILE FOUNDATIONS.

It is this uncertainty of settlement, and limit to the bearing capacity of the clay, which would seem to make the pile the best foundation, if its use can be effected with consistent economy.

Pile foundations were used in Chicago for many years previous to the introduction of the isolated pier method, and some of the oldest and heaviest buildings are founded on them; notably the grain elevators along the Chicago River, which, in spite of their constantly varying loads, have so far maintained their integrity, though few buildings could be more trying on any type of foundations.

Some twenty years ago the use of piles in Chicago was decried in consequence of the very slipshod methods and designs used in the City Hall building. And as we look back upon the results of this work, it is hardly surprising that piles should have been viewed with suspicion for some time after, by those, at least, who looked no deeper than the effect, without considering the cause. In this building the piles were driven so near together that when a new one was driven its neighbor was raised up. The foundations were put in uniformly, although the weight was far from being uniform on the different piers; and even by the time the floors were placed a variation of $7\frac{1}{2}''$ had resulted in the settling.

Another very good example of poor pile-driving at about the same time were the foundations for the Chicago water-works tower. The surface material consisted of about 17' of pure lake-shore sand, and a very heavy hammer was needed to drive a pile even $\frac{1}{4}''$ by measurement, the hammer rebounding three and four times. But the specifications as to depth had to be complied with, and the piles were hammered and hammered until the sand was pierced through, and a drop of 11'' was suddenly noticed.

After these and other failures the stone and concrete foundation was used, until the introduction of the "raft" method, which was almost universally approved, and so extensively used that the pile method was for a time quite dispensed with. But in 1889 Mr. S. S. Beman revived the use of piles in the Wisconsin Central Depot, under trying circumstances. The building itself is only eight stories high, while the tower, carried on piles at 20 tons and more per pile, is 240 ft. high. There has been no appreciable unequal settlement.

Another firm advocate of the pile foundation is Felix Adler of the firm of Adler & Sullivan. The Schiller Theatre Building, by these architects, was built on piles, "as the enormous concentrations of loads, next to adjacent walls, made it seem almost impossible to use iron and concrete foundations without an expense almost prohibitive." So it was decided to use piles, driven 50 ft. below datum, loaded at 55 tons per pile, and cut off at datum, with oak grillage on top and a solid bed of concrete spread over the entire area. Mr. Adler, who is one of the best authorities on pile foundations in Chicago, states as follows on this case:

"As the tendency in pile-driving was to raise the surrounding earth, we watched the adjacent buildings carefully. It was found on driving the piles in the first lot that an adjacent building had settled 6 in., and had to be raised on screws; and throughout the pile-driving these settlements were noticed, requiring the greatest care. Another surprise was that of the four surrounding buildings the one with the least efficient foundations was the only one not requiring such attention, and the piles were driven right up to the building-line without movement of the walls. Under the Borden Block, the heaviest of the adjoining buildings, the movement was such as to require holding up, and inserting new foundations.

"Another peculiarity, which seemed to be a legitimate outcome of the pile-driving, was the apparent readjustment of the particles of clay and sand into the condition of jelly, thus destroying the resisting qualities. The water in the soil is not thoroughly mixed, but occurs in strata or pockets; hence the jar of the driving caused the sand, clay, and water to mix, forming a jelly. The water also rushed into the Schiller site from under the Borden Block, undoubtedly explaining some of the settlements."

These remarks of Mr. Adler certainly show that the work in question was not at all successful as regards the adjacent property, and, indeed, such damage was done by the pile-driving in the case of the Schiller Theatre that suit was instituted against the owners of that building, by the owners of the adjacent Borden Block, as a result of damage sustained. A similar suit was brought against the proprietors of the Stock Exchange building, and the results of the suits now pending must largely settle the foundation question in Chicago. The outcome is awaited with much interest by all of the architects interested in high-building methods.

The new Chicago Library foundations are perhaps the most carefully executed pile foundations in Chicago, being designed and executed by Gen. Sooysmith. Under the walls of this building three rows of piles were driven, and the tests were made as follows: To give the conditions as they would be in the final structure, three rows of piles were driven in a trench, and the middle row was cut off below the other two, thus bringing all the bearing on four piles only (two in each outside row), but thereby allowing the outside rows to derive the benefit of the compression of the earth due to the driving of the central row. The work was done by a Nasmyth hammer, weighing 4500 lbs., falling 42 in., and having a velocity of 54 blows per minute. The last 20 ft. were driven with an oak follower. The piles were driven at $2\frac{1}{2}$ ft. centres to a depth of 52 ft., 27 ft. into soft clay, 23 ft. into hard clay, and 2 ft. into the hard-pan.

Their average diameter was 13 in., and the area at the small end 80 sq. in.

The bearing power of the hard-pan was taken at 200 lbs. per sq. in. Rankine's formula gives about 170 lbs. The extreme average frictional resistance per sq. in. of the sides of the piles, deduced from experiments under analogous conditions, was 15 lbs. per sq. in. The extreme resistance at the pile point was 200 lbs. $\times$ 80 = 1600 lbs. The average external surface of one pile equalled ($52 \times 12 \times 4\frac{1}{3}$) = 25,000 sq. in. At 15 lbs. per sq. in. this gives 375,000 lbs., or $195\frac{1}{2}$ tons. Disregarding the point resistance, the bearing power of a pile would be about 187 tons.

Assuming the ultimate crushing strength of wet Norway pine not over 1,600 lbs. per sq. in., and with a factor of safety of 3, the safe load will be not over 533 lbs. per sq. in. The piles were taken at an average area of 113 sq. in., which gives not over 60,230 lbs. per pile, or about 30 tons. This gives a factor of 3 for crushing, and a factor of 6 for the frictional resistance of the soil. If the timber were loaded at one half its ultimate strength, 45 tons could be used per pile.

A platform to hold a load of pig iron was built resting on the outside rows of piles, and the weight was gradually increased until at the end of eleven days the mass was 38 ft. high, weighing 404,800 lbs. on 4 piles, or about $50\frac{7}{10}$ tons per pile. Levels were taken at intervals of two weeks, and as no settlement was observed, 30 tons per pile was considered a safe load.

Tests were also made of drawing piles at this site, and an ordinary pile, driven in clay to a depth of 45 ft., gave 45,000 lbs. resistance.

In localities where bed-rock itself cannot be reached with economy, piles will undoubtedly give the most satisfactory results, if they can be driven to bed-rock or hard-pan, the tops cut off below the water-line, and all this without damage to surrounding property.

A prominent point in the criticisms of Gen. Sooysmith on Chicago high-building methods is his recommendation of deep piling to bed-rock, with the tops cut off 15 ft. below datum. While this would doubtless be a good thing, it is entirely unnecessary in the opinion of the writer, and far too expensive. Some reasons for this difference of opinion are the following : A number of high buildings supported on piles driven to hard-pan only, with the tops cut off at datum, are proving very satisfactory. Among others may be mentioned the Home Insurance Building, which has settled so uniformly that the greatest variation in levels throughout the whole is but three-fourths of an inch. Piling to bed-rock would necessarily be very expensive in many localities, and in parts of Chicago this would mean 80 ft. below the sidewalk level; and if the piles were driven from a sub-basement, as proposed by Gen. Sooysmith, the trouble and expense of draining this area below the sewer level would be very great. If piles are to be used at all, a proper penetration of the hard dry clay would seem sufficient, with the tops cut off at datum. The large grain elevators along the Chicago River, with their constantly varying loads, which prove a most severe test, have stood without blemish, as before said.

And that such piling is the only system of foundations to be recommended, as Gen. Sooysmith thinks, might be questioned. There can be no doubt that proper piling, or caissons sunk to bed-rock, must be employed where room cannot be had for steel foundations proportioned at 3000 lbs. per sq. ft. of clay area, but some of the disadvantages of piling have already been pointed out. The general law of damage to adjacent property includes the driving of pile foundations, and the difficulty encountered in caring for surrounding buildings must certainly not be overlooked. Where all buildings are built on piles, the adjacent property need not be injured.

Another objection to piling next to buildings supported on steel foundations lies in the difficulty of supporting the walls on screws to allow for additional settlement during and after the placing of the new foundations. This can always be done when new steel foundations are used, but it becomes much more difficult and dangerous with the use of piles.

The method of independent piers and raft foundations has certainly proved quite satisfactory in its very extensive use in Chicago, and, with such uniform settlement as has resulted, on account of the care that was taken beforehand, it answers all the requirements made of it. The writer has a preference for pile foundations, but the many advantages that attend the other kind must be freely acknowledged.

PNEUMATIC FOUNDATIONS.

Pneumatic caissons have lately been employed in a notable example of high building construction in New York City, namely in the Manhattan Life Insurance Building. The building proper is seventeen stories high, with a tower on top, terminating in a dome. The main roof is at an elevation of 242′ 0″ from the widewalk, and from sidewalk to base of flagstaff = 347′ 6″, and from base of foundations to top of dome = 408′ 0″. This makes the dome 61′ 0″ higher than the neighboring spire of "Old Trinity."

The area of the lot is, approximately, 120′ 0″ deep × 67′ 0″ frontage, or 8,000 square feet, which, with the estimated total weight of the building of some 30,000 tons, would give a load of 7,500 lbs. per square foot of lot area.

The natural soil at the site consisted of mud and quicksand to a depth of some 54' 0", down to bed-rock. Had piles been used, as close together as the New York building law allows, or 30" centre to centre, over the entire area, some 1323 piles could have been driven, with an average load of 45,300 lbs. each. This was inadmissible, as the building law limits the load per pile to 40,000 lbs. each, when driven 2' 6" centres.

A new departure in foundations was therefore necessary, especially as the surrounding buildings were built on the natural earth, making them particularly liable to injury in case of any increase of pressure on the soil from additional loading, or decrease in pressure through deep excavations or trenches for piles or concrete piers below the adjacent footings.

Pneumatic caissons were thus adopted, the work being executed by Sooysmith & Co. This was the first example of the pneumatic system as applied to buildings, although the same architects, Kimball & Thompson, had before used smaller caissons in the Fifth Avenue Theatre building in New York City, but without the use of compressed air.

Fifteen caissons, varying in size from 9' 9" in diameter to 25' 0" square, supported the thirty-four cast-iron columns. These caissons were sunk to an average depth of about 31' 6" below the bottoms of the excavations at the site. After the caissons were sunk to bed-rock the rock surface was dressed and stepped as required, and the chambers and shafts were then rammed with concrete, composed of 1 part Alsen cement, 2 parts sand, 4 parts broken stone. The superimposed piers were built of hard-burned brick laid in cement mortar. About eight days were required to sink each caisson. The locations of the several caissons are shown in Fig. 119.

A very elaborate system of cantilever girders was used to transfer the loads on the columns in the side walls to

FOUNDATIONS.

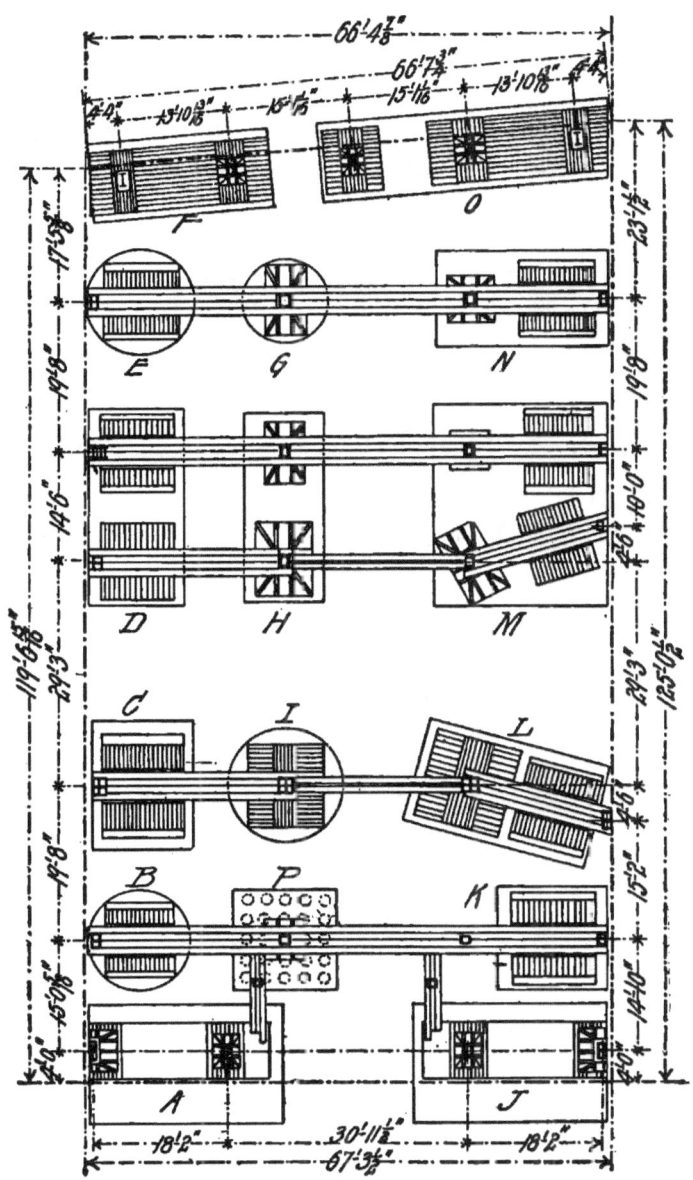

Fig. 119.

proper concentric bearings over the caisson piers. From these bearings the load was distributed over the whole masonry-work by means of large steel bolsters, thus

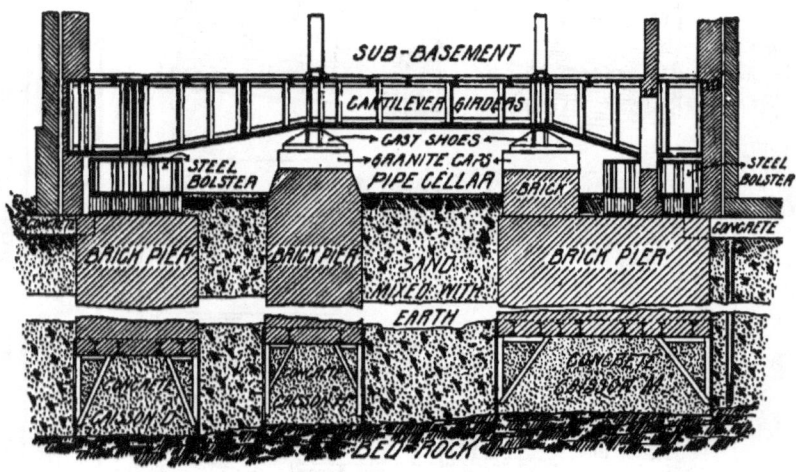

FIG. 120.

diminishing and equalizing the unit-pressure. A cross-section of the caissons and cantilever girders is shown in Fig. 120.

CHAPTER XI.

UNIT-STRAINS—SPECIFICATIONS.

THE question of unit-strains will naturally vary to a considerable extent with the personal opinions of the designer—the more conservative his views the lower his allowances. But, whatever the preferences of the engineer or architect, he is, to a large measure, limited by the city building laws with which he is required to conform. A comparison between the building ordinances of New York, Chicago, and Boston, given in the next chapter, will show the wide divergence which exists in their respective requirements.

A few unit-strains will here be mentioned as having been employed in Chicago skeleton buildings before the adoption of the present ordinance. Cast iron and timber will not be considered as entering into modern high-building construction.

BRICKWORK.

The allowable pressure per square foot on brick masonry as used in the highest masonry piers in Chicago, namely, in the Masonic Temple, has been mentioned before as 12 tons.

Prof. I. O. Baker, in his "Treatise on Masonry Construction," gives the following allowable strains on brickwork as the practice of the leading Chicago architects:

10 tons per sq. ft. on best brickwork laid in 1 to 2 Portland cement mortar;

8 tons per sq. ft. for good brick laid in 1 to 2 Rosendale cement mortar;

5 tons per sq. ft. for ordinary brick, laid in lime mortar.

He shows, however, that these figures are very conservative, as his tables of the ultimate strength of best brickwork give from 110 tons with lime mortar to 180 tons with Portland cement mortar per square foot. So while the 12 tons in the Masonic Temple was even greater than ordinary Chicago practice, Prof. Baker adds that " reasonably good brick laid in lime mortar should be safe under a pressure of 20 tons per sq. ft."

COLUMNS.

We have few experiments of value on the ultimate strength of full-sized columns of the type most used at present. Building operations have to be conducted too quickly to allow many tests on the full-sized columns before using. Tests have been made on the full-sized Gray columns, and on the Larimer column, as before referred to. The only full-sized tests on Z-bar columns were made by C. L. Strobel, then Chief Engineer of the Keystone Bridge Company (see Transactions of the American Society of Engineers, April, 1888), who introduced this shape into the United States. But even these tests are hardly fair ones for present comparisons, as lattice bars were used instead of web plates, and almost all the tests were for a much higher ratio of the radius of gyration to the length of column than is ordinarily met with in building work. It seems as though higher breaking loads would be obtained for the majority of columns as used at the present time. Burr, in his " Strength and Resistance of Materials," deduces formulæ for the Keystone and Phœnix columns, but none for the Z column or the box column of plates and angles. The latter type was used in the Masonic Temple in two-story lengths, lattice bars being used instead of the plates in the lighter columns. But as the height of a single story was less than 12' 0" unsupported length, a uniform unit-strain of 12,500 lbs. per sq. in. was used without reduction by the radius of gyration, for all concentric loading. Columns with eccentric loads were figured for a unit-strain of 12,500 lbs. per sq. in. reduced by Rankine's formula for eccentric loading.

In the Venetian Building the columns without strains from wind bracing were figured at 15,000 lbs. per sq. in. for all concentric dead and live loads, with an extra allowance for eccentric loads. The columns carrying strains from the wind bracing were figured at 20,000 lbs. per sq. in. for all concentric loads,—dead, live, and wind,—with an additional allowance for eccentric loading. In these columns the wind-strains amounted to from 35 to 40 per cent of the total load, so that this mode of treatment of using a higher unit-strain gave a much greater section to the column than if a lower unit-strain had been used and the wind forces disregarded. These unit-strains have been used in a number of Chicago high buildings, notwithstanding some rather severe criticism.

In "The Fair" Building, by W. L. B. Jenney, architect, 12,000 lbs. was used uniformly on all columns, with no allowance for eccentric loading. This building is one of the heaviest in the city of Chicago, being figured for 130 lbs. live load per square foot for the 1st, 2d, 3d, 4th, and 6th floors, 200 lbs. for the 5th floor, 100 lbs. for the 7th and 8th floors, with the rest at 75 lbs., all in addition to dead loads. Great care was taken in providing good connections throughout.

In the Fort Dearborn Building, by the same architect, a uniform unit-strain of 13,000 lbs. per sq. in. was used on all columns, made of channels and plates, with a proper reduction for eccentric loading.

The writer believes that with the use of a mild steel, of an ultimate strength of from 65,000 to 68,000 lbs. per sq. in., 15,000 or 16,000 lbs. per sq. in. may safely be used for all concentric dead, live, and wind loads combined (with an additional allowance for eccentric loading as before described), provided that the wind pressure is taken at not less than 30 lbs. per sq. ft., and that the live loads on the floor systems are assumed as required by the municipal building laws. With careful regard for all connections, and remembering that the strength of a structure lies in its weakest point, these unit-strains would seem to satisfy both the conditions of proper economy and satisfactory design.

The use of 20,000 lbs. per sq. in., as in the Venetian Building, would seem too high, especially when the live load is but 35 lbs. per sq. ft. on the floor systems, and when but 50 per cent of this is considered as transferred to the columns.

SPECIFICATIONS FOR STRUCTURAL STEELWORK.

Material and Workmanship.—The entire structural framework, as indicated by the framing plans, or specified, is to be of wrought steel, of quality hereinafter designated, all material to be provided and put in place by this contractor unless specifically stated to the contrary. All work to be done in a neat and skilful manner, as per detail or specified, and if not detailed or specified, as directed by the superintendent to his entire satisfaction.

Quality and Material.—Steel may be made by either the Bessemer or open hearth process. It shall be uniform in quality, and must not in any case contain over 0.10 of 1 per cent of phosphorus.

The grade of steel used (except for rivets) shall fill the following requirements when tested in small specimens:

Ultimate tensile strength: 60,000 to 68,000 lbs. per sq. in.
Elastic limit: Not less than one half the ultimate strength.
Elongation: Not less than 20 per cent in 8 in.
Reduction in area: Not less than 40 per cent at point of fracture.

Bending Test.—Duplicate specimens will be required to stand bending 180° around a mandrel, the diameter of which is equal to one and a half times the thickness of the specimen, without showing signs of rupture on either concave or convex side. After being heated to a dark cherry red, and quenched in water at 180° Fahr., the specimen must stand bending as before.

Inspection.—All steelwork is to be inspected front the melt to final delivery of finished material on board cars. The inspection will include surface, mill, and shop inspection by an inspector satisfactory to the engineer, to whom all reports are to be made. No work shall be delivered until approved and stamped by the inspector. All inspection shall be at the expense of this contractor.

Tests.—A test from the finished material will be required representing each blow or cast. In case the blows or casts from which the blooms, slabs, or billets in any reheating furnace charge are taken, have been tested, a test representing the furnace heat will be required, and must conform to the requirements as before specified.

The original blow or cast number must be stamped on each ingot from said blow or cast, and this same number, together with the furnace heat number, must be stamped on each piece of the finished material from said blow, cast, or furnace heat.

Rivet Steel.—The steel used for rivets shall fulfil the following requirements:

Ultimate tensile strength : 56,000 to 62,000 lbs. per sq. in.
Elastic limit : Not less than 30,000 lbs. per sq. in.
Elongation : Not less than 25 per cent in 8 in.
Reduction of area at point of fracture shall be at least 50 per cent.

Specimens from the original bar must stand bending 180° and close down on themselves without sign of fracture on convex side of curve. Specimens must stand cold hammering to one third the original thickness without flaying or cracking, and must stand quenching as heretofore required for rolled specimens.

Cast Iron.—All cast iron shall be of the best quality of metal for the purpose intended. Castings shall be clean and free from defects of every kind, and boldly filleted at all angles.

The cast iron must stand the following test:

A bar 1" square, 5' 0" long, 4' 6" between bearings, shall support a centre load of 550 lbs. without sign of fracture.

Drawings.—All copies of architects' drawings, shop drawings, templates, patterns, models, etc., and all necessary measurements at the building, shall be made by this contractor at his own expense. All shop drawings must be submitted for the approval of the architects, and such changes or additions shall be made as are required by said architects or their agent.

Painting.—No material shall be painted until approved by the inspector, nor shall any painting be done when material is exposed to rain, or in otherwise improper condition. No material shall be shipped until the paint is thoroughly dry.

All iron and steel shall receive one coat of best red lead ground in linseed-oil before leaving the shop. When the framework is completed, all exposed portions are to be touched up with paint as specified, and the whole shall then receive a second coat of best red lead mixed with linseed-oil.

Beams.—All floor, roof, and other beams shown on framing plans to be of size and weight shown, and accurately located according to plan. Where two or more beams are shown side by side, they shall be provided with cast separators at least every 8' 0" apart, but with never less than three in each span. Each separator to be at least ¾" thick, and cast to fit the profile of the beam exactly. Separators must be provided at each and every bearing. Where the distance centre to centre of beams is not given in the drawings, they shall be set at the minimum distance given in Carnegie's table of separators.

Girders.—All plate and lattice girders to be proportioned to the following stresses per square inch:
Extreme fibre stress.... 12,000 lbs.
Compression. .... 10,000 "
Tension................ 12,000 "
Shearing............... 6,000 " for webs, 9,000 for rivets.
Direct bearing, including rivets, 15,000 lbs.

In all built girders the flanges alone are to be considered as resisting the bending moments. Both flanges to be of the same section, the net section to be figured in all cases. No angles to be used smaller than $2\frac{1}{2}'' \times 2\frac{1}{2}'' \times \frac{5}{16}''$, and no webs to be of a thickness less than ⅜". Wherever the distance between the flange-angles is greater than 70 times the thickness of the web, stiffening angles shall be used not farther apart than the total depth of the girder. Stiffeners must be provided at all bearings and at points of concentrated loading. All stiffeners to be placed over filler plates, and ends of stiffeners, top and bottom, to fit closely against the flange-angles.

Columns.—The maximum strain upon the metal in columns shall not exceed 12,000 lbs. per sq. in. for a length less than or equal to 90 radii of gyration. For columns of a greater length the metal shall be proportioned by the formula $17,000 - \frac{60l}{r}$, l to be taken in inches. No column to have an unsupported length of more than 30 times its least lateral dimension. The least radius of gyration shall be used.

All columns, where possible, shall be made in two-story lengths, breaking joints alternately. Columns to be built with *vertical* connection-plates or splice-plates, all joints to be equal in strength to the column itself. Bearing-surfaces must be "finished" and protected by white lead and tallow. All columns must be perfectly true and tested at frequent intervals. "Shimming" will not be allowed.

Castings.—Cast iron used in the shape of lintels, corbels, or brackets shall be so proportioned that the compressive strain does not exceed 13,500 lbs. per sq. in., nor the tensile strain exceed 3,000 lbs. per sq. in. Cast-iron plates may be loaded to 15,000 lbs. per sq. in. Cast-iron column bases may be strained to 6,000 lbs. fibre strain. They shall not give a pressure of more than 15 tons per sq. ft. on brickwork, nor more than 30 tons per sq. ft. on granite.

Plates.—Cast-iron plates shall be set under ends of all beams and girders, resting on masonry, so proportioned as not to exceed a load of 15 tons per sq. ft. on brickwork, nor more than 30 tons per sq. ft. on stone.

Connections—Splices.—All field-connections and splices to be riveted with hot rivets. Where girders or beams rest on brackets attached to the columns, such beams or girders shall be riveted through the bottom flanges to the bracket, and also have connection-angles connecting the top flanges to the column. The ends of all girders or beams resting on masonry walls or piers to have anchors securely embedded in the masonry-work.

Rivets.—All rivets to be of mild steel, as before specified. The pitch of rivets shall never be less than $1\frac{1}{2}''$ nor more than $6''$, while the minimum distance from the centre of any rivet to the edge of material shall be $1\frac{1}{4}''$. No rivets to be used in tension. An excess of 25 per cent shall be allowed in proportioning field-rivets. Rivet-holes may be punched or drilled, but must not be more than $\frac{1}{16}''$ larger than diameter of rivet. Rivet-holes must be accurately spaced, as drift-pins will be allowed for assembling only. The rivets shall completely fill the holes, with full heads concentric with the rivets, and in full contact with the surface of the material.

SPECIFICATIONS FOR BRICKWORK, ETC.

(Extracts from Masonry Specifications for the Fort Dearborn Building. Jenney & Mundie, Architects.)

This contractor will furnish and set all that part colored red on the drawings, and not shown or specified for pressed brick or terra-cotta; to be the best character of common brickwork, laid up with the best merchantable, good, sound hard bricks, acceptable to the architects, to lines and levels on all sides, in lime mortar, all joints being carefully filled and the bricks rubbed well into place and pounded down to make a small solid joint. When laid in dry, warm weather, bricks will be laid wet. The joints of all outside common brick, and of all interior brickwork not to be plastered, shall be neatly struck and cleaned down.

Pressed-brick Work.—The contractor will furnish and set all that part colored red on the drawings and marked or shown to be pressed-brick work, to include all returns into openings, with the best character of pressed-brick facing of even color and of the kind and character hereinafter specified. All exposed brickwork of areas and entrances in fronts marked to be finished in pressed-brick work shall be faced with the same character of pressed brick as used in the adjacent parts. All joints in the pressed-brick work to be neatly rodded. All pressed-brick work to be laid from an outside scaffold in mortar the color of the brick. All courses to be gauged true. In laying pressed brick each edge and down the middle is to be buttered and all vertical joints to be filled from front to back. The returns of pressed-brick work must be carefully dovetailed into the common brickwork or banded by solid headers.

In the piers only solid headers must be used. A sample of pressed brick is to be deposited with the architects.

This contractor will furnish and set the terra-cotta or salt-glazed tile copings to all masonry walls not covered by stone or metal copings. The copings are to be 2 inches wider than the wall and to have lapped joints. Copings to be set in Portland cement.

Concrete.—This contractor will furnish and set all concrete foundations or concrete filling shown on the drawings. All concrete shall consist of equal parts of Portland cement, mortar, and broken stone. The size of broken stone is to be that of small egg coal. The mortar is to be thoroughly mixed, and the stone to be wet before mixing with mortar. The concrete to be cut over twice. No more water to be used than is necessary to moisten every particle of cement. All concrete to be used immediately after mixing, and shall be pounded hard in place until the water stands on the top of the concrete.

Cement Plastering.—The outside of all masonry walls that will come in contact with the earth shall be smooth plastered by this contractor with a surface coat of Portland cement mortar of an average thickness of $\frac{1}{2}$ inch from the lower footings to the top of finished grade.

Protection.—This contractor will carefully protect his work by all necessary bracing, and by covering up all walls at night, in bad weather, and at all times when work is liable to be interrupted either by storms or cold. He will protect all masonry-work from frosts by covering with manure or other material satisfactory to the architects. The top of all walls injured by the weather shall be taken down by this contractor at his expense before recommencing work.

Footings.—Concrete footings shall be enclosed by 2-inch plank curb, said plank to be left in place. All water is to be baled out of trenches before the concrete is put in.

SPECIFICATIONS FOR FIRE-PROOFING.

(Extracts from Fire-proofing Specifications for the Fort Dearborn Building. Jenney & Mundie, Architects, Chicago.)

The following specifications include the fire-proofing of all the steel in the building, the filling in between the beams forming floors, and the concreting over the same to the top of the floor-strips, and the projections of the beams below the arches.

Also the covering of all columns, both those standing clear and those partly incased in the walls.

Also the building of all tile partitions and the tile vaults. Also the building of the party-walls over the present old brick walls. Also the tile floor of the roof and pent-houses on the roof.

All work shall be laid in mortar composed of 3 parts of best fresh lime mortar and 1 part best Louisville cement, thoroughly mixed together at time of using. Said lime mortar shall be composed of fresh burned lime and clean sharp sand in proportions best suited to this work.

This contractor shall furnish all material, including the mortar for setting the same, and will do all his own hoisting and set all the work in a thoroughly substantial and workmanlike manner to the satisfaction of the superintendent.

Floors.—All floors shall be supported on flat arches set in between the beams and of a shape that shall give a uniform flat ceiling in the rooms below.

The bottoms and projections of all beams and girders shall be protected by projecting parts of tile or by separate beam slabs. In laying the floor arches every joint shall be filled full over its entire surface, from top to bottom.

Floor arches, ten days after they are laid and before they are concreted, shall stand a test of a roller, 15 inches face, and loaded so as to weigh 1500 pounds, rolled over them in any direction.

All columns shall be covered with column tile held by metal clamps both in horizontal and vertical joints. These column protections shall be so made as to conform with the city ordinance.

Roof.—The roof shall be supported in the same way as the floors, only the soffits may be segmental.

Partitions.—All the partitions shown in the several plans are to be built including all cross and subdivision partitions. All are to be of hollow tile 4 inches thick in the first and second stories, and 3 inches thick in all other stories for cross-partitions. All hall partitions to be 4 inches thick.

In glazed partitions the lower parts and all parts other than the sash and frames shall be of tile.

The tiles shall be set breaking joints, and be tied with metal ties or clamps.

All vaults shown on plans above second story to be built with vestibules, as shown.

Furring.—The outside walls in the basement, in the part for rent, will be furred with 3-inch tile, so as to form a vertical and true surface for plastering.

All tilework shall be straight and true.

All tilework shall be thoroughly burned and free from serious cracks or checks or other damages, and shall be laid in a proper and workmanlike manner.

No centres to be lowered until the mortar has set hard.

All structural steel on which the strength of the building depends in any way, including wind bracing, shall be protected by fire-proof covering.

Concreting.—This contractor shall fill in on top of the tile arches with dry cinder concrete, composed of reasonably clean soft-coal cinders, to be levelled off at the top of the highest beams or girders, and after the floor-strips are set to be filled in between said strips with said dry cinders, pressed down hard and leaving a surface reasonably uniform $\frac{1}{8}$ inch below the tops of the strips, so that the floor can be laid without disturbing the cinders.

All damages to tilework to be repaired before the cinders are laid.

Party-walls.—Above the present walls on the west and south sides this contractor shall furnish and lay in the aforesaid cement and lime mortar hard-burnt wall tile. Said wall to be composed of two 6-inch tile between columns, and elsewhere three thicknesses of 4-inch tile clamped together both in the length and across the wall. The face of the outside tile shall be guaranteed to stand weather for five years, dating from the completion of said wall; the contractor agreeing to replace any tile injured by the weather either in winter or summer during said period.

Every joint in this wall, both vertical and horizontal, shall be thoroughly filled over its entire surface with the mortar before mentioned. All outside joints to be struck in a neat and workmanlike manner.

TERRA-COTTA SPECIFICATIONS.

Material.—This contractor shall furnish and set wherever called for on drawings terra-cotta to exactly match in color the sample submitted, all in strict accordance with detail drawings. Material for all terra-cotta to be carefully selected clay, left in perfect condition after burning, and uniform in color. All pieces to be perfectly straight and true, and with mould of uniform size where continuous. No warped or discolored pieces will be allowed. This contractor to furnish a sufficient number of over-pieces, so as to avoid all delay.

Modelling.—All work shall be carefully modelled by skilled workmen, in strict accordance to detail drawings, and models shall be submitted for architects' approval before work is burned. No work burnt without such approval will be accepted by the architects unless perfectly satisfactory.

Mortar.—All mortar used for exposed joints in terra-cotta work shall correspond in every particular with mortar used for pressed-brick work. It shall be composed of lime putty, colored with "Pecora" or "Peerless" mortar stains; colors to be selected by the architects.

Ornamental Fronts, Belt Courses, Bands.—This contractor shall furnish and set all ornamental terra-cotta, belt courses, and bands, as shown on elevations or sections, or where otherwise indicated, in strict accordance with detail drawings. All terra-cotta work to be secured to the iron-work in the most approved manner, with substantial wrought-iron or copper anchors, and thoroughly bedded in cement mortar. All horizontal courses to have lap joints. All projecting courses to have drips formed on the under side.

Caps and Jambs, Sills.—All caps and jambs where indicated as terra-cotta will be constructed in strict accordance with detail drawings. All sills and belt courses to have counter-sunk cement joints as directed by the superintendent. All projecting sills to have drips formed on under side, and all sills shall be raggled for hoop iron, which shall be bedded by this contractor in cement mortar.

Terra-cotta Mullions.—All ornamental mullions of terra-cotta to be secured to metal uprights in approved manner, and well bedded and slushed with cement mortar.

Cornice.—This contractor shall construct cornice in strict accordance with detail drawings, with sufficient projection through walls and approved anchorage to the metal-work to make same thoroughly secure, this contractor to furnish all necessary anchors. Form raggle in cornice as shown for connection of gutter, this raggle to be on face of terra-cotta. Leave openings in cornice for down-spouts as shown.

Anchors.—This contractor shall furnish all anchors, of substantial wrought iron or copper, for the proper support and anchoring of all terra-cotta used in his work. All terra-cotta to be drawn to tight and accurate joints, to entire satisfaction of the superintendent. All terra-cotta must fit the supporting metal-work exactly.

Cutting and Fitting.—This contractor shall do all cutting and fitting necessary to make his work perfect in every particular, all possible cutting and fitting to be done at the factory before delivery.

Protection of Terra-cotta.—All projecting terra-cotta shall be protected with sound plank during the erection of the building by terra-cotta contractor, said protection pieces to be removed on cleaning down of building.

Cleaning Down.—This contractor shall carefully clean down all terra-cotta work at completion of building, when directed by the superintendent, and shall carefully point up all joints before leaving work.

CHAPTER XII.

BUILDING LAWS.

THE building ordinances of the cities of New York, Chicago, and Boston are all of comparatively recent adoption, and though perhaps no one of them may lay claim to being a model building law, still one might expect to find much of the best practice and experience in building construction incorporated in one or all of these laws.

Some of the more important subjects coming under the head of Architectural Engineering may be compared as follows:

FLOOR LOADS.

The requirements for live loads per square foot on the floor-beams, over and above the dead weight of the floor itself, are:

	New York.	Chicago.	Boston.
1. Dwellings (*a*)	70	70 ⎫	70
2. Office buildings (*b*)	100	70 ⎬ (*e*)	100
3. Public buildings (*c*)	120	70 ⎭	150
4. Stores, warehouses, factories, etc. (*d*)	150 and upward.	150 minimum. Posted notices of Allowable Load.	250 Posted notices in Bldgs. for Mechanical or Mercantile Purposes.

(*a*) Includes hotels and apartments in New York.
(*b*) Includes apartments and boarding- and lodging-houses in Chicago.
(*c*) Called "places of public assembly" in New York.
(*d*) Includes stables in Chicago.
(*e*) Allowances may be made for reduction in these loads on columns and foundations.

It will be seen that these three laws agree in a live load of 70 lbs. per square foot for private dwellings. This is undoubtedly high, 40 lbs. per square foot being about the average in use by the best engineers and consulting architects. This requirement, taken with the value given for the strength of wooden beams in the Boston law, necessitates timbers of far larger size than has been the practice of the best architects, or as used in houses which have been built and occupied from thirty to fifty years. Kidder shows that *actual* loads in parlors (including piano), dining-rooms, etc., average only 14 to 23 lbs. per square foot of the whole area. The excessiveness of the load of 70 lbs. for dwellings would seem to be further indicated by the use of the same load in the Chicago laws for classes 2 and 3. New York and Boston are about alike for these two classes; but if 70 lbs. is sufficient for office and public buildings, why require it for lighter private dwellings?

In class 4 each city law requires that all floors for warehouses, etc., must be carefully computed, according to the intended use, and the capacity of such floors be posted in conspicuous places about the building. The New York and Chicago laws are much more explicit on this point than is the Boston law, while the Chicago ordinance leaves the required load to the judgment of the architect or engineer with the approval of the Building Commissioner. The minimum load of 150 lbs. in the New York law is far too small in many cases, but the loads for these types of buildings are hard to classify, and are best left to the care of competent designers under the approval of the building departments. Mr. W. L. B. Jenney had occasion to estimate the loads in the wholesale warehouse of Marshall Field & Co. in Chicago, and the surprisingly low average of 50 lbs. per square foot was found for the total floor area, including all passage-ways. The maximum load on limited areas was found to be 57 lbs.

The writer sees no reason for changing the previous recommendations of live loads, as given under a discussion of the floor system, namely :

40 lbs. per square foot for dwellings;

80 to 90 lbs. for places of public gatherings, devotional, educational, or amusement;

40 lbs. for the upper floors of office buildings;

80 lbs. for the lower floors of office buildings;

and from 150 to 450 lbs. for places of manufacture, storage, machinery, etc.

WROUGHT IRON: STRESSES IN POUNDS PER SQUARE INCH.

	New York.	Chicago.	Boston.
Extreme fibre stress, rolled beams and shapes..................	12,000	11,000 beams or rails in foundations. 12,000	12,000
Tension.......................		12,000	12,000
Compression in flanges, built beams......................	12,000	10,000	10,000
Shearing.....................	9,000 rivets. 6,000 webs.	7,500 shop rivets 6,000 field rivets 6,000 webs.	9,000
Direct bearing, including pins and rivets...................	15,000		15,000
Bending on pins..............			18,000
Modulus of elasticity..........			27,000,000

STEEL: STRESSES IN POUNDS PER SQUARE INCH.

	New York.	Chicago.	Boston.
Extreme fibre stress, rolled beams and shapes..................	15,000	14,000 in foundations. 16,000	16,000
Tension.......................		16,000	15,000
Compression in flanges, built beams......................	15,000	13,500	12,000
Shearing.....................	9,000 rivets. 7,000 webs.	9,000 shop rivets 7,500 field rivets. 10,000 webs.	10,000
Direct bearing, including pins and rivets...................	15,000		18,000
Bending on pins..............			22,500
Modulus of elasticity.........			29,000,000

As may be seen from these tables, the Boston law is the most comprehensive, while the Chicago ordinance is singularly deficient in unit-stresses, and even somewhat contradictory in some of the few values given. Thus under the heading of plate girders, fibre stresses of 13,500 lbs. per square inch for steel and 10,000 lbs. for wrought iron are allowed, while in a preceding section "all girders, beams, corbels, brackets, and trusses" are allowed fibre stresses of 16,000 lbs. for steel, and 12,000 lbs. for wrought iron. This latter section does not limit the use of these unit-stresses to either rolled or built members, thus clashing with the requirements for plate girders. Still different fibre stresses are called for under the requirements for rail or beam foundations, 14,000 lbs. per square inch for steel, and 11,000 lbs. for wrought iron. No values are given for bearing.

The New York law, under the provisions for plate girders, specifies that "no part of the web shall be estimated as flange area, *nor more than ½ of that portion of the angle-iron which lies against the web.*" As the effective depth of the girder is limited to the distance between the centres of gravity of the flange areas, this requirement would seem quite unnecessary. If the web be neglected as affecting the flange area, and proper deductions made for rivet-holes, the whole angle areas can very properly be used.

COLUMNS.

The New York and Boston laws both call for computations by Gordon's formula, using the constants of 12,000 lbs. per square inch for steel, and 10,000 lbs. for wrought-iron columns. No column is to have an unsupported length of more than 30 times its least lateral dimension, nor to have metal less than $\frac{1}{4}''$ in thickness.

The Chicago law allows the use of the constant of 12,000 lbs. per square inch for wrought-iron columns, or

$$S = 12,000a \div \left(1 + \frac{l^2}{36,000r^2}\right) a, l, \text{ and } r \text{ all in inches.}$$

For steel columns two formulæ are given:

$S = 17,000 - \left(\frac{60l}{r}\right)$ for columns more than 60 radii in length, and $S = 13,500a$ for columns under 60 radii in length (l and r both in inches). The formula for columns over 60 radii in length gives about 13,000 lbs. per square inch for a column in which $\frac{l}{r} = 66$.

CAST COLUMNS.

The New York law specifies that the computations for cast columns shall be made by the use of Gordon's formula, with the constant of 16,000 lbs. per square inch. All cast columns to have a minimum average thickness of ¾″, with an unsupported length of not more than 20 times their least lateral dimensions.

The Chicago law gives formulæ for both round and rectangular cast columns. For round cast columns:

$S = 10,000a \div \left(1 + \frac{l^2}{600d^2}\right)$ l = length of column in in.; d = diam. of columns in in.; a = sectional area col. in in.

For rectangular cast columns:

$S = 10,000a \div \left(1 + \frac{l^2}{800d^2}\right)$ l and a as before; d = the side of square column, or the least horizontal dimension of other columns.

The Boston law provides tables for both round and square cast columns.

STONE.

The use of Stone for Walls, Facings, Piers, Arches, etc., is thus Specified, in Tons per Square Foot.

	New York.	Chicago.	Boston.
Granite............		⅒ of the ultimate strength derived from tests, approved by Commissioner of Buildings. Dressed stone, laid in Portland cement.	60 ⎫ First quality,
Marble and limestone..	Not specified.		40 ⎬ dressed beds, laid solid in
Sandstone...........			30 ⎭ cement mortar.

The safe loads given in the Boston law are about double those recommended by Baker, while the Chicago requirements, using $\frac{1}{30}$ of the average ultimate strengths given by Prof. Baker, allow 38 tons on granite, 30 tons on limestone, and 24 tons on sandstone per square foot.

The use of ashlar masonry in wall facings is limited as follows: Boston law: "In reckoning the thickness of walls ashlar shall not be included unless it be at least 8" thick. In walls required to be 16" thick or over the full thickness of the ashlar shall be allowed; in walls less than 16" thick, only half the thickness of the ashlar shall be included. Ashlar shall be at least 4" thick, and properly held by metal clamps to the backing, or properly bonded to the same."

Chicago law: "Stone may be used as facing for brick walls under the following conditions: If the facing is ashlar, without bond courses, and the individual courses thereof measure in height between bond stones more than six times the thickness of the ashlar, then each piece of ashlar facing shall be united to the brickwork with iron anchors, at least two to each piece, and reaching at least 8" over the brick wall, and hooked into the stone facing as well as the brick backing. Wherever ashlar, as before described, is used, it shall not be counted as forming part of the bearing-surface of the wall, and the brick backing shall be of the thickness of wall herein specified for the different kinds of building.

"If stone facing is used with bond courses at a distance apart of not more than six times the thickness of the ashlar, and where the width of bearing of the bond courses upon the backing of such ashlar is at least twice the thickness of the ashlar, and in no case less than 8", then such ashlar facing shall be counted as forming part of the wall, and the total thickness of wall and facing shall not be required to be more than herein specified for walls of the different classes of buildings."

New York law: "All stone used for the facing of any building, and known as ashlar, shall not be less than 4" thick.

Stone ashlar shall be anchored to the backing, and the backing shall be of such thickness as to make the walls (independent of the ashlar) conform, as to the thickness, with the requirements of this ordinance."

Dimension stones, as specified in the Chicago ordinance for foundations, shall not be subjected to a load of more than 10 tons per square foot. If the beds of the stones are dressed and levelled off to uniform surface, and the stones are set in Portland cement mortar, this load may be increased to 25 tons per square foot.

BRICKWORK: ALLOWABLE PRESSURES IN TONS PER SQUARE FOOT.

	New York.	Chicago.	Boston.
Brickwork laid in cement mortar..................	15 ⎫	15 tons with Portland cement. ⎫	15 ⎫
Brickwork laid in cement and lime mortar.........	11½ ⎬ (a)	12 tons with ordinary cement. ⎬ (b)	12 ⎬ (c)
Brickwork laid in lime mortar..................	8 ⎭	8 tons with lime mortar. ⎭	8 ⎭

(*a*) Isolated brick piers shall not exceed 12 times their least dimensions.
(*b*) The loads permitted for brick piers shall be 25 per cent less than in walls.
In walls an additional 25 per cent may be allowed if brickwork is thoroughly grouted or "shoved."
A further 20 per cent additional allowance may be made if walls are built of sewer brick only, or, if vitrified paving bricks are used, this allowance may be made 30 per cent.
(*c*) In brick piers in which the height is from 6 to 12 times the least dimension, these pressures are reduced to 13, 10, and 7 tons respectively for the mortars as above given.

BEARING POWER OF PILES AND SOILS.
Given in Pounds per Pile, or per Square Foot on Footings.

	New York.	Chicago.	Boston.
Piles............................	40,000	50,000	
Pure clay, at least 15 ft. thick...		4,000	
Dry sand, at least 15 ft. thick...	Not specified.	3,500	Not specified.
Clay and sand, mixture..........		3,000	
"Good solid, natural earth"...	8,000		

It would certainly seem quite remarkable that a city of the size of Boston should fail to specify any unit loads for foundations. For ordinary footings the only requirements are that "the foundation, with the superstructure which it supports, shall not overload the material on which it rests." Piles are specified for use "where the nature of the ground requires it," and " the number, diameter, and bearing of such piles shall be sufficient to support the superstructure proposed." All piles must be capped with block granite levellers, and must not be over 3' 0" centres in the direction of the wall.

It is to be hoped that the building department of the city of Boston is less of a political organization than is the case in most large American cities, or that the contractors of that city are more conscientious than the average. The New York laws, in the requirements for pile foundations, permit a 5" point, while no mention is made of the butt end. Piles are usually specified with 8" points and 14" butts, and for such piles the allowable pressure of 20 tons under the New York law is certainly very conservative; 30 tons are very commonly used for piles of 6" to 8" point, and 12" to 16" butt.

The New York laws also require that "if, in place of a continuous foundation wall, isolated piers are to be built to support the superstructure where the nature of the ground and the character of the building make it necessary, inverted arches shall be turned between the piers at least 12" thick and of the full width of the piers." This practice has long since been condemned in Chicago, as in no way satisfactory or desirable.

The Chicago building ordinance is certainly far superior to those we have just mentioned, as regards the subject of foundations, and the following quotations would seem to recommend themselves for general application.

"Foundations shall be proportioned to the *actual* average

loads they will have to carry in the completed and occupied building, and not to *theoretical* or *occasional* loads."

" Foundations shall be constructed of either of the following: Portland cement concrete, or Portland cement concrete and steel or iron, or dimension stone, or sewer or paving brick, or timber piles covered with grillage of oak timber, or a grillage of oak timber alone; it being, however, provided that timber shall not be used in connection with any foundation at a level higher than city datum."

"Where pile foundations are used, borings of the same shall first be made to determine the position of the underlying stratum of hard clay or rock, and the piles shall be made long enough to reach to hard clay or rock, and they shall be driven down to reach the same, and such piles shall not be loaded more than 25 tons to each pile. The heads of the piles are to be protected against splitting while they are being driven, and after having been driven the piles are to be sawed off to uniform level and covered with an oak timber grillage, so proportioned that in the transmission of strains from pile to pile the extreme fibre strain in the timbers composing the grillage shall not be more than 1200 lbs. to the square inch."

The bearings on other materials than piles are then given, as in previous table. The cement to be used in concrete footings "shall not be less than 90 per cent fine on 80-mesh sieve, and when mixed one part of cement to one part of clean, sharp sand, moulded into briquettes of one square inch cross-section, shall not break when seven days old at less than 225 lbs. tensile strain, nor at thirty days at less than 275 lbs. tensile strain."

In view of the many discussions at the present day it will be interesting to note the requirements for the coating or painting of rails or beams in foundations.

The Boston law requires that "all metal foundations and

all constructional ironwork underground shall be protected from dampness by concrete, in addition to two coats of red lead, or other material approved by the inspector."

New York law: "When crib footings of iron or steel are used below the water-level, the same shall be entirely coated with coal-tar, paraffine varnish, or other suitable preparation before being placed in position. When footings of iron or steel for columns are placed below the water-level, they shall be similarly coated for preservation against rust."

The Chicago ordinance requires a perfect covering of concrete only: "If steel or iron rails or beams are used as parts of foundations, they must be thoroughly embedded in a concrete, the ingredients of which must be such that after proper ramming the interior of the mass will be free from cavities. The beams or rails must be entirely enveloped in concrete, and around the exposed external surfaces of such concrete foundations there must be a coating of Portland cement mortar not less than one inch thick.

WIND PRESSURE.

No mention is made of the wind pressure to be figured in either the New York or Boston law, except that the former law requires a live load of 50 lbs. per square foot to be taken for all roofs.

The Chicago law provides as follows: "In the case of all buildings the height of which is more than $1\frac{1}{2}$ times their least horizontal dimension, allowances shall be made for wind pressure, which shall not be figured at less than 30 lbs. for each square foot of exposed surface. The precautions against the effects of wind pressure may take the form of any one or all of the following factors of resistance to wind pressure:

"First. Dead weight of structure, especially in its lower parts.

"Second. Diagonal braces.

"Third. Rigidity of connections between vertical and horizontal members.

"Fourth. By constructing iron or steel pillars in such manner as to pass through two stories with joints breaking in alternate stories."

ALLOWABLE HEIGHT OF BUILDINGS.

The New York law sets no limitation on the height of buildings in that city.

Boston law: "No building or other structure hereafter erected, except a church spire, shall be of a height exceeding $2\frac{1}{2}$ times the width of the widest street on which the building or structure stands, whether such street is a public street or place or a private way existing at the passage of this act or thereafter approved as provided by law, nor exceeding 125 feet in any case; such width to be the width from the face of the building or structure to the line of the street on the other side, or if the street is of uneven width, such width to be the average width of the part of the street opposite the building or structure."

Chicago ordinance: "No building shall be erected in the city of Chicago of greater height than 160 feet from the sidewalk level to the highest point of external bearing walls. And the height of no building of skeleton construction shall be more than three times its least horizontal dimension. And no building of masonry construction shall be more than four times as high as its least horizontal dimension."

The buildings which have been termed "sky-scrapers" in Chicago were all built before the passage of this ordinance, or on building permits which were issued before the law went into effect.

APPENDIX TABLE—GIVING THE CONSTRUCTIVE DATA OF THE PRINCIPAL OFFICE BUILDINGS OF CHICAGO.

Architects.	Building.	Area.	Height.	No. of Stories.	Kind of Columns.	Type of Floors.	Partitions.	Exterior Walls.	Wind Bracing.	Foundations.
W. L. B. Jenney	Home Insurance	14,500 sq. ft.	164' 0"	12	Cast; Keystone used in 2 additional stories	Tile	Tile	Veneer	Cross-walls in court	Stone on 18" of concrete.
"	Manhattan	10,040 "	196' 10"	16			"	"	Rods	Rails and beams on 12" of concrete.
"	"The Fair"	32,000 "	136' 0"	9	Z bar		"	"	"	Rails and beams on 12" of concrete.
"	Leiter	54,870 "	133' 6"	8	Cast	Tile	"	"	None	Beams on 16" of concrete.
"	Y. M. C. A.	14,200 "	251' 0" tower	13	Z bar	"	"	"	Rods	Beams on 16" of concrete.
"	Isabella	3,420 "	166' 0" to ridge	10		"	"	"	"	Rails and beams on 12" of concrete.
Jenney & Mundie	New York Life	11,300 "	156' 4" to coping	12	Box column of plates and angles	"	"	"	Channels in exterior walls	Beams on 12" of concrete.
"	Fort Dearborn	6,400 "	190' 0"	12	Channels and plates	Soft tile	Soft tile	"	Gusset-plate knee-bracing in the exterior walls	Beams on 18" of concrete.
Holabird & Roche	Tacoma	8,140 "	166' 0"	13	Cast	Tile	Tile	"	Cross-walls	Beams on concrete.
"	Pontiac	5,340 "	174' 6"	14	Z bar	"	"	"	"	"
"	Caxton	5,340 "	150' 6"	12	"	"	"	"	"	"
"	Venetian	5,522 "	157' 0"	13	"	"	"	"	Rods	"
"	Katahdin Wachusetts Monadnock new.	13,267 "	203' 6"	17	"	"	"	Solid walls, part veneer	"	"
"	Old Colony	10,115 "	213' 0"	17	Z bar and Phoenix	"	Plaster board	Veneer	"	"
"	Champlain	7,070 "	189' 0"	15	Z bar	"	Tile	"	"	"
"	Marquette	24,190 "	207' 0"	16	"	"	"	"	"	Rails and beams on concrete
Adler & Sullivan	Auditorium	53,726 "	Bldg. = 146' Tow'r = 265' Bldg. = 10 Tow'r = 17		Cast	"	"	Solid	Walls only	Timber grille, rails, and beams.
"	Schiller Theatre	14,400 "	Bldg. = 112' Tow'r = 212' Bldg. = 13 Tow'r = 17		Z bar and Phoenix	Tile; some concrete arches used	Mackolite	Part solid " veneer	Girders	Piles, timber grille, beams.
"	Stock Exchange	18,000 "	173' 0"	13	Z bar		Tile	Veneer	None	Piles and beams.
Burnham & Root	Rookery	29,760 sq. ft.	164' 0"	12	Cast	Hard tile	"	Solid	None—solid walls only	Rails and beams.

APPENDIX TABLE—GIVING THE CONSTUCTIVE DATA OF THE PRINCIPAL OFFICE BUILDINGS OF CHICAGO.

Architects.	Building.	Area.	Height.	No. of Stories.	Kind of Columns.	Type of Floors.	Partitions.	Exterior Walls.	Wind Bracing.	Foundations.
Burnham & Root	Monadnock	28,000 sq. ft.	215′ 0″	16	Z bar	Porous tile	Mackolite	Solid	Solid walls—portal-struts, flats in floors	Rails and beams
" "	Great Northern Hotel	16,500 "	168′ 0″	14	"	Porous tile	Tile	Veneer	Rods	Rails and beams
" "	Phenix	20,825 "	160′ 0″	11	Cast	Hard tile	"	Solid	Walls only	Rails.
" "	Woman's Temple	18,184 "	196′ 5″	13	Z bar	Johnson's arch	"	"	"	Rails and beams
" "	Masonic Temple	19,224 "	273′ 0″	20	Plates and angles	Hard tile, Johnson New	"	Walls carry themselves only	Rods	" "
" "	Ashland	11,260 "	200′ 7″	16	Z bar	hard tile, Johnson New	"	Veneer	None	Rails.
" "	Marshall Field	16,459 "	153′ 0″	9	"	hard tile, Johnson	"	Solid	Walls only	Rails and beams
" "	Rand-McNally	....	120′ 0″	10	"	Tile	"	Veneer	Horizontal rods	" "
D. H. Burnham & Co.	Reliance	4,675 "	200′ 0″	14	Gray	Porous tile	Porous tile	"	Plate girders in exterior walls	Dimension stone
Henry Ives Cobb	Newberry Library	21,000 "	100′ 0″	5	Larimer Phoenix	Tile	Tile	Solid	None	Rails and concrete,
" "	Title and Trust	11,000 "	197′ 5″	16				Veneer	Bracing	Demension stone and concrete.
" "	Owings	3,650 "	170′ 0″	14	Cast	"	"	Solid	None	Beams and concrete.
" "	Chicago Athletic Association	13,686 "	144′ 0″	10	Z bar	"	"	"	"	" "
" "	Boyce	3,673 "	149′ 0″	12	"	"	"	Veneer	Bracing	Plate girders and concrete.
" "	Hartford	4,710 "	173′ 0″	14	Cast	"	"	"	"	Beams and concrete.
Clinton J. Warren	Unity	....	210′ 0″	17	Plates and angles	"	"	"	Rods	Beams and [crete.
" "	Security	4,650 "	200′ 0″	14	Cast	"	"	Support themselves	....	Beams and concrete.
" "	Auditorium Annex	....	....	10	Z bar	"	"	"	"	Beams and [crete.
W. W. Boyington	Columbus	....	....	......	Plates and channels	Terra-cotta lumber	Terra-cotta lumber	Veneer	Knee-bracing in exterior walls	Beams and concrete.
Handy & Cady	Teutonic	4,860 "	136′ 6″	10	"	"	"	"	"	Rails and beams on concrete.

INDEX.

	PAGE
Anchors for terra-cotta work	104
specifications for	215
Ashland Block, data about	228
wind-bracing in	148
Athletic Club Building, data about	228
fire in	13, 75
fire-proofing of columns	20
Auditorium, data about	227
foundations of	190
settlement of	191
Auditorium Annex, data about	228
Bay windows, construction of	107
floors and ceilings in	112
framing of, for Reliance Building	109
Masonic Temple	109
spandrel sections, Reliance Building	112
Beam footings, calculation of	180
foundations	178
Beams in floor system	82
spandrel	100
specifications for	206
Book-tile	164
Borings for foundations	171
Boston Building Law—allowable height of buildings	226
floor loads	216
foundations	222
loads on brick-work	222
loads on masonry	220
strength of columns	219, 220
wrought-iron and steel	218
Box columns	124, 126, 131
fire-proofing of	134
Boyce Building, data about	228
Brackets for bay windows	108
Reliance Building	112

INDEX.

	PAGE
Brick, hollow, used for fire-proofing	134
Brick-work—allowable pressure on	201
building laws	220
specifications for	209
Building Laws	216
brick-work	222
cast columns	220
columns	219
foundations	222
height of buildings	226
stone, walls, piers, etc.	220
wind pressure	225
wrought-iron and steel	218
Built sections *vs.* rolled beams	160
Caissons—pneumatic	198
Cantilever girders	186, 198
Cast columns—building laws	220
disadvantages of	114
joints for	114
Castings—specifications for	208
Cast-iron—specifications for	206
Caxton Building—data about	227
Ceilings—suspended	165
Cement plastering—specifications for	210
Champlain Building—data about	227
floor-plan of	39
Chicago Building Law—allowable height of buildings	226
fire-proof construction defined	11
fire-proofing of exterior columns	95
interior columns	135
floor arches	74
floor loads	216
foundations	222
loads on brick-work	222
masonry	220
mill construction defined	18
skeleton construction defined	94
slow-burning construction defined	16
strength of columns	219, 220
wind pressure	225
wrought-iron and steel	218
Chicago Construction	91
Chicago Library—pile foundations	194
Chicago Office Buildings—data about	227, 228
Chicago Stock Exchange—data about	227
description of	26
Clearance between floor-beams, girders and columns	85
Columbus Building—data about	228

INDEX. 231

	PAGE
Column-brackets for bay windows	112
-connections	127
Pabst Building, Milwaukee	159
-formula	117
-joints	156
-loads, in Fort Dearborn Building	82
-loads in "The Fair" Building	177
-sheets	168
Columns—building laws	219
capabilities of fire-proofing	129
cast *vs.* wrought	113
choice of	131
cost of	119
details in Venetian Building	145
eccentric loading	124
examples of great length	181
expansion and contraction of	97
fire-proofing of	132
Athletic Club Building	20
Chicago Law	135
New York Law	96
Gray type	125
joints for cast-iron	114
Larimer type	121
limestone pillars	131
patent	119
Phœnix type with pintle-plates	125
placed in exterior walls	89
plates and angles	124
practical considerations	119
principles of resistance	116
requirements for fire-proofing	133
riveting of	121
Schiller Theater Building	118
shopwork and workmanship	120
specifications for	207
splices in Reliance Building	159
tabulation of loads	169
theoretical form	116
two-story lengths	158
types of, for building work	115
unit strains on	202
used for pipe-space	129
vertical splices	157
Z-bar sections used in "The Fair" Building	130
Y. M. C. A. Building	130
type	123
Combined footings	185

	PAGE
Combined footings, calculation of	186
Compression of clay under foundations	176
Concrete floor-arches	66, 69, 71
in foundations	184
specifications for	210
Connection-angles—standard	85
Connections—specifications for	208
Court walls	107
Dead loads	76
on floor system	80
of Fort Dearborn Building	82
on foundations	176
Deflection of floor-beams	84
framework, due to wind	160
Detail plans for steelwork	43
Drawings—specifications for	206
Earthquakes—provisions for	156
Eccentric loading—calculation of	126
on columns	124
Elevator enclosures	166
Erection of steel-work	46
cost of	46
cranes used in	46
time required for	46
Field connections	46
Fire loss in United States	9
Fire-proof construction—comparative cost of	10
definition of	11
Fire-proof ducts for piping	21
structures—requirements of	16
vaults	165
Fire-proofing, efficiency of	131
in Tremont Temple, Boston	22
materials for	12
methods of	15
of columns	20, 132
of stairways and elevator shafts	19
specifications for	211
Fire test of fire-proof building	13
Floor arches, brick	54
Chicago Building Law for	74
comparative costs of	74
corrugated iron	54
Guastavino type	73
hollow tile	54
in Equitable Building	56
in Home Insurance Building	56
in Montauk Building	56

INDEX.

	PAGE
Floor arches, Melan system	67
segmental	72
steel straps and concrete	68
test by fire	75
test of Metropolitan system	70
wire mesh	69
Floor-beams, calculation of	84
Chicago practice	82
connections for	85
deflection of	84
economical arrangement of	83
necessary clearance	85
Floor-girders	86
length of	86
Floor-loads	76
Fort Dearborn Building	81
Marshall Field Building	80
Old Colony Building	81
requirements of Building Laws	216
"The Fair" Building	177
Floors, specifications for	211
Fort Dearborn Building—data about	227
description of	38
floor and column loads	81, 82
unit strains on columns	203
wind-bracing	155
Foundations	171
Auditorium	190
beam	178
Building Laws	222
calculation of beam footings	180
combined footings	186
rail footings	179
Chicago Library	194
combined footings	185
concrete in	184
Great Northern Hotel	179
independent piers	173
loads on	176
Manhattan Building	186
Life Insurance Building, New York	197
Marquette Building	184
masonry *vs.* raft	173
Old Colony Building	178
pile	192
pile *vs.* raft	196
pneumatic	197
rail	178

	PAGE
Foundations, Rand-McNally Building	186
Schiller Theater Building	193
settlements of	191
"The Fair" Building	177
Wisconsin Central Depot	193
Framing plans	39
economical	83
Furring, specifications for	212
tile	165
Girder loads—Fort Dearborn Building	82
Girders, cantilever	186, 198
for floor system	86
spandrel	101
specifications for	207
Gray column, details of	125
Great Northern Hotel, data about	228
, foundations of	179
Guastavino floor arches	73
Hartford Building, data about	228
Height of buildings—building laws	226
Hollow tile	54
advantages of	15
floor arches	56
sustaining power	62
used for furring	165
used in partitions	163
Home Insurance Building	97
data about	227
settlement of	191
Inspection, specifications for	205
Iron, wrought, building laws	218
Isabella Building, data about	227
wind-bracing	154
Jackscrews used in foundations	185
Johnson's patent tile-arch	61
Joints, open	90
Knee-braces, calculation of	153
Knee-bracing—Fort Dearborn Building	155
Isabella Building	154
Larimer columns—connections	121
tests of	123
Lee tile-arches	58
Leiter Building, data about	227
Lime vs. cement	50
Limestone pillars vs. steel columns	131
Live loads—Chicago practice	79
defined	76
discussion of, for office buildings	77

	PAGE
Live loads on foundations	176
in Mills Building, San Francisco	79
in Venetian Building	79
Manhattan Building, data about	227
foundations of	186
Life Insurance Building, New York, foundations of	197
Marquette Building, data about	227
description of	26
foundations of	184
Marshall Field Building, data about	228
floor loads	80
Masonic Temple, box columns in	124
column-sheets in	169
data about	228
mechanical plants in	33
piers in	90
special features in	36
two-story columns in	158
unit strains on columns	202
wind-bracing in	143
Masonry—building laws	220
piers	88
Mechanical features, installation of	21
Melan floor-arches	67
Metropolitan floor-arches	69
Mill construction	18
Monadnock Building, data about	227, 228
settlement of	191
vibrations due to wind	161
wind-bracing in	152
Mortar, colored	214
Mullions, connections of	101
specifications for	214
Newberry Library, data about	228
New York Building Laws—fire-proofing of columns	96
floor loads	216
foundations	222
loads on brick-work	222
masonry	220
protection of steel-work	53
strength of columns	219, 220
wrought-iron and steel	218
New York Life Insurance Building, data about	227
description of	38
time required for erection	46
Old Colony Building—column connections	128
data about	227
floor loads	81

	PAGE
Old Colony Building—foundations	178
wind-bracing	152
Owings Building, data about	228
Pabst Building—column connections	159
Painting of metal work, specifications for	206
Panelled beams	58
Partitions—load per square feet on floor system	80
specifications for	212
types of	163
used in wind-bracing	136
Permanency of skeleton construction	50
Phœnix Building, data about	228
columns, connections of	128
fire-proofing of	134
with pintle-plates	125
Piers—exterior—Chicago type	91
Marshall Field Building	89
Masonic Temple	90
Monadnock Building	98
treatment of	88
Pile foundations	192
Chicago Library	194
tests of	194
Piles—building laws	222
Pioneer tile-arches	58
Plaster used as fire-proofing	135
Plates, specifications for	208
Pneumatic caissons	198
foundations	197
Pontiac Building, data about	227
deflections due to wind	162
Porous tile in fire-proofing	134
Portal bracing, calculation of	148
Old Colony Building	152
Poulson floor-arches	72
Pressed-brick work, specifications for	209
Rail footings, calculation of	179
foundations	178
Rails, properties of	179
Rand-McNally Building, data about	228
foundations of	186
Reliance Building, data about	228
description of	26
splices in columns	159
wind-bracing of	156
Rivets, specifications for	208
steel, specifications for	205
Rods for wind-bracing	148

	PAGE
Roof construction	164
Roofs, specifications for	212
Rookery Building, data about	227
Schiller Theater Building, data about	227
foundations of	193
Security Building, data about	228
Segmental floor-arches	72
Separators	86
Settlement, allowance for	172
Chicago Post Office	172
of exterior walls	89
of foundations	191
use of jackscrews in	185
Skeleton construction, defined	94
earliest example of	96
permanency of	50
Skew-backs in tile-arches	57, 58
Slow-burning construction	16
Spandrel sections—Ashland Block	101
bay windows	108
for Reliance Building	112
Fort Dearborn Building	101
Marshall Field Building	107
Marquette Building	105
Masonic Temple, bay windows	109
through court walls	107
Spandrels, defined	100
Specifications for brick-work	209
fire-proofing	211
structural steel-work	204
terra-cotta	213
Stairways	166
Steel—requirements of building laws	218
Steel-work, deterioration of	50
in walls, protection of	95
painting of	53
protection of	51
Boston law	53
Chicago law	53
New York law	53
specifications for	204
time required for erection	46
with cement mortar	52
with lime mortar	50
Stone—building laws	220
Struts—wind-bracing in Venetian Building	147
Sway-rods, calculation of, for wind-pressure	140
typical calculation of	143

	PAGE
Tacoma Building, data about	227
Terra-cotta, anchors for	104
enamelled	16
for exterior walls	91
specifications for	213
used for column fire-proofing	133
Tests of steel-work, specifications for	205
Teutonic Building, data about	228
"The Fair" Building, data about	227
foundations	177
floor loads	177
loads on columns	177
settlement of	191
unit strains on columns	203
wind-bracing in	144
Tie-rods for floor-arches	62
Tile-arches for roofs	164
necessary tests for	63
tests of	59
types most used	61
weights of	59
Tile floor-arches, construction of	56
Tile floors, calculation of	64
Tile, hard *vs.* porous	133
Title and Trust Building, data about	228
Tremont Temple, Boston, fire-proofing in	22
Unit strains	201
on columns	202
Unity Building, data about	228
erection of steel-work	46
Vaults, fire-proof	165
Veneer construction	102
Venetian Building, column sheets in	169
data about	227
floor loads	79
unit strains on columns	203
wind-bracing	144
Walls, allowable pressure on	201
compression of	89
exterior	88
Chicago type	91
thickness of	98
with spandrel girders	101
court	107
settlement of exterior	89
solid masonry, objections to	88
Western Bank-note Building, settlement of	191
Wind-bracing—calculation of knee-braces	153

	PAGE
Wind-bracing—calculation of portal-bracing	148
sway-rods	141
Chicago practice	137
diversity of practice in	136
in Ashland Block	148
Fort Dearborn Building	155
Isabella Building	154
Masonic Temple	143
Monadnock Building	152
Old Colony Building	152
Reliance Building	156
"The Fair" Building	144
Venetian Building	144
types of	139
Wind-pressure—building laws	225
calculation of sway-rods	140
experiments on deflection	161
Fort Dearborn Building	155
limiting height of building	160
practical considerations	140
unit loads	138
World's Columbian Exposition	138
Wisconsin Central Depot, foundations of	193
Woman's Temple, data about	228
foundations of	173
World's Columbian Exposition—wind-pressure	138
Y. M. C. A. Building, data about	227
Z bar columns, fire-proofing of	134
Monadnock Building	134
objections to	123
large sections of	130

www.ingramcontent.com/pod-product-compliance
Lightning Source LLC
Chambersburg PA
CBHW020759230426
43666CB00007B/773